MARC ZIMMER, PhD

GLOWING GENES
A Revolution in Biotechnology

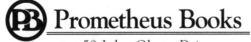 **Prometheus Books**

59 John Glenn Drive
Amherst, New York 14228-2197

Published 2005 by Prometheus Books

Inquiries should be addressed to
Prometheus Books
59 John Glenn Drive
Amherst, New York 14228–2197
VOICE: 716–691–0133, ext. 207
FAX: 716–564–2711
WWW.PROMETHEUSBOOKS.COM

09 08 07 06 05   5 4 3 2 1

Library of Congress Cataloging-in-Publication Data

Zimmer, Marc.
    Glowing genes : a revolution in biotechnology / Marc Zimmer.
        p. ; cm.
    Includes bibliographical references and index.
    ISBN 1–59102–253–3 (hardcover : alk. paper)
    1. Green fluorescent protein—Popular works. [DNLM: 1. Luminescent Proteins—genetics—Popular Works. 2. Genetic Engineering—Popular Works. 3. Luminescence—Popular Works. 4. Organisms, Genetically Modified—Popular Works. QU 55 Z72g 2005] I. Title.

QP552.G73Z56 2005
572'.4358—dc22

2004022817

Printed in the United States on acid-free paper

CONTENTS

ACKNOWLEDGMENTS

I owe many people thanks. My wife, Dianne, and my children, Matthew and Caitlin, who always smile and listen when I start talking about science. My parents, thanks for instilling an interest and love for nature in me and for everything you have done. Ms. Munting and Ms. Hofmeyer (Sasolburg High School), Prof. Rob Hancock (Wits), Prof. Nick Kildahl (W.P.I.), Prof. Bob Crabtree (Yale), and Prof. Gene Gallagher (Connecticut College) have been my mentors and role models; without them I might have been a tax consultant. Drs. Shimomura, Chalfie, Prasher, Ward, Wood, Hoffman, and Branchini for taking the time and talking to me about "glowing genes." The cancer chapter was written for you, Viola. Thanks, Vicki Fontneau, for enthusiastically reading *The Rise and Fall of Nicholas Evans*, the novel that still needs to be published. Everyone who let me use their glowing gene photographs, thanks and sorry I couldn't use them all. The print shop folks—thanks for your interest in my pictures and writings. My agent, Ed Knappman, and my editors, Linda Greenspan and Heather Ammermuller, helped bring this book to life. The National Institute of Health, the Research Corporation, and the Camille and Henry Dreyfus Foundation have funded my green fluorescent protein research, which provided the impetus for this book. I owe a debt of gratitude to my colleagues in the Department of Chemistry and at Connecticut College for making it fun to be a chemistry geek. The Rossi-Reders, Gallaghers, and the makers of red wine, *boerewors*, and biltong all have a special place in my heart. Antjie Krog, Andre Brink, Bill Bryson, and Wally Lamb—I wish I could write like you have. Thanks to all my students— you make me feel young and inspire me to find articles about glowing pig snouts and glowing sperm. Finally, thank *you* for picking up and reading this book.

INTRODUCTION

Take some jaded grown-ups canoeing around Vieques Island, Puerto Rico, where every paddling motion generates thousands of bursts of light, and you will see the same enthusiasm shown by children when they chase fireflies or adorn glow-in-the-dark necklaces. (See figure 1 in the photo insert.) This fascination of ours with living organisms that produce light is what has led us to a revolution in biotechnology.

The scientists who started this revolution had no intention of doing so. They were interested in doing basic science. Their prime concern was to establish how and why some living organisms produce light. In an attempt to answer that question and similar ones, they tediously extracted the chemicals that are responsible for producing light from bioluminescent organisms like fireflies and jellyfish. They wanted to understand the chemistry responsible for producing light in bioluminescent organisms and never dreamed that they were pioneering the way for future scientists to use these glowing proteins in medicine, art, the defense industry, and agriculture. The story of glowing genes is a great example of how the quest for basic science has quite unexpectedly produced some extremely practical and useful techniques.

"Glowing genes" have lit up a whole new world. They are the microscopes of the new millennium. Let's look at bacteria, an extremely difficult task, because they are minuscule. If you wanted to use your naked eyes to see a bacterium sitting on the period at the end of the last sentence, you would have to enlarge the period so that it was at least one hundred feet across.[1] Now imagine trying to see where that bacterium is inside a living mouse. Well, by using firefly luciferase, the same chemical that fireflies use to produce light, one can see bacteria inside a live mouse and follow them around. A new generation of microscopes has been developed that can measure the amount of light given off so that you can even determine how many bacteria there are by simply measuring the amount of light that is produced in a given location. One way of examining the efficacy of a particular antibiotic is to inject a lab rat with bacteria that have been made luminescent through firefly genes. It is then fairly simple to monitor the spread of the infection by observing how the luminescence spreads throughout the mouse. If the antibiotic is effective, then, after it has been administered, the infection will recede, and the luminescence will diminish.

Although there are thousands of organisms that emit light, glowing gene technology to date has only used proteins from jellyfish, fireflies, bacteria, sea pansies, and corals. The glowing genes from these creatures have been used to monitor and image biological processes in genetics, transplantation biology, gene therapy, molecular biology, cellular biology, and medicinal, food, environmental, and military diagnostics.

Although we have just been introduced to luciferase—the protein responsible for the light flashes emitted by fireflies—it is the green fluorescent protein from the jellyfish, *Aequorea victoria*, that is the most commonly used glowing protein. In fact, it is so commonly used that one can even talk of a green fluorescent protein (GFP) revolution.

Jellyfish have lit up the oceans of the world for the last 650 million years. Two thousand years ago, Pliny the Elder first described a use for the luminescence found in some jellyfish. He found that you could take certain jellyfish and smear them against a walking stick to make it glow so that it would light up the path on an evening stroll. We can think of this as the first seeds of the GFP revolution. Flash forward to 1994, when scientist Martin Chalfie showed that it was possible to take the green fluorescent protein gene out of the jellyfish and place it in any other cell where it could be made to glow. Within five years, GFP could be found in most molecular biology laboratories throughout the world. Rabbits, zebra fish, cancer cells, and grapevines were all fluorescing, thanks to GFP.

Proteins are even smaller than bacteria. If you wanted to be able to see a protein in a period, the protein and period would have to be expanded so that the period was at least a mile across. The reason GFP is so useful is because it can be fused to the ends of proteins, making them fluoresce without changing their ability to function or move through the cell. This has enabled scientists, using specialized equipment, to observe when proteins are made in living organisms and how they move. It has opened a window to a world that we have never been able to see before, but it also makes us consider the ethical ramifications of the rapid advances being made in genetic engineering. Thanks to GFP's fluorescence, cancer researchers can now see when a cancer cell metastasizes and moves to another part of a model organism. Developmental biologists can take one of sixty-four cells from a developing embryo and label all the proteins made from that cell and its offspring. This would be impossible without GFP.

A literature search for references that had green fluorescent protein in the title and/or abstract found only ten publications released in all of 1994, while an identical search in February 2004 found 133 research papers that were published in that month alone. The 1994 paper by Professor Chalfie that started the GFP revolution is among the twenty most-cited papers in the field of molecular biology and genetics.

Although more than nine thousand papers have been written about this fascinating protein, very few people know anything about the history of GFP technology, and no one knows why jellyfish fluoresce.

Glowing genes have greatly reduced the need for the use of radioisotopes and for sacrificing laboratory animals. They light up biological processes without interfering with them, which has made them incredibly useful and thus has triggered their widespread use in research. However, in order to use green fluorescent protein and all the other glowing genes in medicine, agriculture, and biotechnology, it is necessary to create transgenic organisms. Many people have morally and philosophically objected to humans changing the genetic makeup of other species. On the one hand, one can defend the creation of genetically modified organisms by arguing that mankind has been cross-breeding plants for hundreds of years, thereby creating genetically modified plants, but then again, the process of actually taking the genetic material from one species and placing it into the genetic material of a completely different species is much more invasive. In this book, I will not try to answer these questions. However, I would love to show you the surprising and truly amazing things that can be done with glowing genes. Perhaps you will be fascinated and even alarmed, or both, by the transgenic organisms that I describe in this book, but it is my hope that you will find these pages instructive and interesting.

I should also note at the outset that many people oppose genetic engineering since they are worried about the safety of genetically modified organisms. Are these animals and plants safe to eat? Can the foreign genes move from plants to weeds and create new superweeds and so forth? Glowing genes have been used to examine the possibility that foreign inserted genes might move from a "transgenic" or genetically modified species to another; this will be discussed later.

Every day, forty thousand people die from malnutrition: at least half of these are children under the age of five. The World Health Organization has estimated that the population of the world will increase from six billion people today to nine billion people by 2050. This means that the production of food has to be doubled while using less water and about the same amount of land.[2] Genetically modified plants, such as the potato plant that glows when it needs to be watered, may have to be part of the solution of this overwhelming problem.

Another advance is that inexpensive arsenic detectors have been created using glowing genes from both the firefly and jellyfish. Arsenic has had a long and sordid history, starting long before the Bangladeshi water contamination crisis. Many a historic murder has been carried out using arsenic. Madame Toffana of Sicily was perhaps the most egregious of the arsenic murderers. She was a seventeenth-century cosmetician with a wicked streak. Today, it is fashionable to have a tanned complexion, and many of us are prepared to take the risk of future skin cancer to look good and healthy. In the seventeenth century, Victorian ladies were horrified at the idea that their red cheeks could result in someone mistaking them for sunburned peasants. Aqua Toffana was the answer; when applied to the face of a socially conscious Victorian lady, it removed any sign of sun exposure. Unfortunately, it was just as effective at removing unwanted spouses, rivals, and lovers, for Aqua Toffana was made from the synovial fluid of the joints of freshly slaughtered pigs that had been rubbed with a solution of arsenic. Aqua Toffana wasn't some gimmick—it really worked. The arsenic in the solution killed the red blood cells on the surface of the skin, resulting in a paler, whiter look. It was also a very effective poison, and it is estimated that as many as six hundred people may have succumbed by unknowingly ingesting some of Madame Toffana's aqua. In case you are worried, Madame Toffana did not go unpunished— she was sentenced to death by public strangulation for her deeds.[3]

According to the World Health Organization, the arsenic poisoning currently occurring in Bangladesh is the world's largest mass poisoning of a population in history. Tens of millions of people, perhaps even as many as half of the country's 137 million, are drinking water that contains unsafe levels of arsenic. The long-term effects of this arsenic poisoning are skin diseases and cancers. The relatively recent increase in arsenic concentration of drinking and irrigation water in Bangladesh and northern India is a very unfortunate side effect of an attempt to help the people of Bangladesh. In the 1970s, many international aid organizations dug millions of wells to provide the Bangladeshi population with a cleaner source of water than the surface water they were using. The project was partially successful, as 97 percent of the population now get their water from wells and no longer have to drink water contaminated with bacteria. The incidence of cholera, diarrhea, and infant mortality subsequently dropped. But many of the wells contain minerals that were slowly dissolving and releasing dangerously high levels of arsenic. By 1993 the first signs of arsenic poisoning were noticed among the population.[4]

The source of the arsenic is almost certainly sedimentary rock laid down over millennia by the rivers, such as the Ganges, that run down from the Himalayas. One of the major problems in dealing with this environmental crisis is that existing tests for arsenic are unreliable, particularly at low but still dangerous concentrations of arsenic. Glowing genes from the firefly and jellyfish are currently being used to develop cheap and accurate arsenic sensors, which are actually genetically modified bacteria. These bacteria, *Escherichia coli*, are commonly found in mammalian guts, nowhere near the Bangladeshi wells. However, researchers have added genes for bioluminescent proteins from both the firefly and the jellyfish into the DNA regions that control arsenic resistance in *Escherichia coli*. When the bacteria are exposed to arsenic, the arsenic resistance genes are activated, and the bacteria begin to glow. The transgenic bacteria have been dried onto paper strips to create inexpensive arsenic-detecting dipsticks.[5] A major problem prevents the widespread use of these "arsenic canaries." *E. coli* is sensitive to many other heavy metals; for example, copper, which is commonly found in the vicinity of arsenic, can kill the bacteria. The lack of light emission caused by the heavy metals that have killed all the modified bacteria from the dead bacteria could then be misinterpreted as an absence of arsenic.

This is just one of many examples of how glowing genes can be used that will be explored in this book. After reading it, you will understand why green fluorescent proteins and luciferases have gone from relatively unknown molecules to ones that are used by thousands of scientists every day. Through these pages, I have not only attempted to describe how glowing genes are transforming scientific research but also tried to provide an insight into the lives of those brilliant pioneers and their successors who have helped to bring forth the glowing-gene revolution.

CHAPTER 1

LIVING LIGHT

What is the most common form of communication on the Earth? Scientists think it may be light signals that are emitted by plants and animals, or, in other words, *bioluminescence*. Many species perform intricate courtship rituals with the light they give off: their glow allows them to recognize members of their own species. It also enables them to attract prey and confuse attackers. The bioluminescent organisms probably first emerged in the "Cambrian explosion," which is when the first species with eyes appear in the fossil record, for surely there would be no reason to give off light if there was nothing else to see it.[1] Most flashing species have their homes in the ocean, where one can find luminescent squid and octopi, glowing jellyfish and shrimp, as well as luminous fish and bacteria. They have descriptive names like the benttooth bristlemouth, the bloodybelly comb jelly, and the flashlight fish. It has been estimated that 60 to 80 percent of all deep-sea fish are bioluminescent.

In some cases, species that seem bioluminescent aren't bioluminescent at all. The light that they emit comes from bacteria living inside them. The flashlight fish is an excellent example; it has a special organ housing bacteria that continuously emit light. It also has a little shutter so that it can "switch" on and off its light source as desired. Some crabs, fishes, and worms have external bioluminescence. They squirt out two chemicals that produce light outside the animal when they mix. However, most animals have internal mechanisms for producing light.

The bristlemouth has two rows of bioluminescent organs on the underside of its stomach so that when predators look up they do not see the bristlemouth's outline against the lighter background. This use of luminescence as "camouflage" is quite common among deep-sea creatures.[2]

The lanternfish live between two thousand and three thousand feet below sea level but will come to the surface at night to find food. They are able to multitask; their luminescence serves as a pattern that is species specific and is used as a recognition device for mating and also for camouflaging them from predators.

The anglerfish is a rather ugly-looking creature that has evolved a unique method for catching its meals. The female has a long fin sticking out in front of her mouth; at the tip of the fin are millions of light-producing bacteria. The fin acts as a fishing rod, while the bacterial luminescence is the bait. Male angler fish grow to a mere two and a half inches long, while the females reach a whopping forty-seven inches. After sex the male bites the female and attaches himself to her. He hangs onto her for dear life, all life. He is completely dependent on the female, has no "fishing rod," and slowly loses his sight once he is attached to his mate. The male connects to the female's bloodstream, and, in exchange for his sperm, he gets protection and food. Some females aren't content with having one hanger-on and have as many as three males attached to them.

The shining tubeshoulder fish lives 650 to 3,000 feet below sea level, and it uses its luminescence as a defense mechanism. When threatened, it can release some luminescent slime from special organs located on its shoulders; the glow distracts predators and allows the shining tubeshoulder fish to escape into the darkness.[3]

Interestingly some species are bioluminescent in certain geographic locations, but not in others. Perhaps this is due to some dietary requirements for bioluminescence that are not present in all areas.[4]

Ninety-seven percent of all the water on the Earth is found in the seas. The average depth of the ocean floor is 2.4 miles, which means that in most areas of these oceans, bioluminescent organisms are the only source of light. Blue light travels the farthest in seawater, so it should come as no surprise that most of the marine bioluminescent creatures give off blue light and can detect only short wavelength radiation, that is, green, blue, indigo, and violet light.

At least seven hundred genera have some bioluminescent species, and there are probably many more deep-sea creatures that have yet to be observed and classified. Bill Bryson has written a great book titled *A Short History of Nearly Everything*, and in it he has a very interesting chapter about water and oceanographic research.[5] He shows how little we really know about the deep seas and its inhabitants. Oceanography was born in 1872, when the British warship *HMS Challenger* undertook a three-and-a-half-year voyage around the world, dredging, measuring, and sampling. More than forty-seven hundred new marine organisms were collected and a fifty-volume report that took nineteen years to prepare was released. The data collected on the trip was very important to scientists of the time, but the trip must have been excruciatingly boring. The repetitive nature of the data collection drove off a quarter of the crew from the ship before she even completed her mission.[6]

In the 1930s, William Beebe and Otis Barton took their search for new species of sea creatures to the depths of the oceans. They were colorful characters, especially Beebe. He was a flamboyant naturalist who shared his adventures with as many people as possible; he wrote articles and books and gave frequent radio broadcasts. He also had a series of attractive female assistants with interesting job descriptions such as "historian and technician" and "assistant in fish problems," which eventually led to a very bitter divorce.[7] So bitter that the *New York Times* used the headline "Naturalist Was Cruel" in its report of the breakup.[8] Like Pliny the Elder and Athanasius Kircher, whom we shall meet in the following chapter, Beebe was also fascinated by volcanoes and nearly died during a volcanic eruption on the Galapagos Island. In the 1920s, he grew frustrated by the number of deep-sea creatures that did not survive being captured in his trawling nets, and he started looking for new ways to observe them in their own habitat. At the same time, Otis Barton, a wealthy engineer, was designing a sphere that could take a person or two down into the unexplored realms of the ocean. William Beebe and Otis Barton met and took oceanography a step further by building the first submersible, which they called a *bathysphere*. They used it to get a close-up look at what was happening far beneath the ocean's surface. In June 1930, in the Bahamas, they became the first people to reach a depth of six hundred feet, and just four years later they reached a depth of 3,028 feet. At that depth, the bathysphere was exposed to pressures of over nineteen tons per square inch. If the steel cable holding the submersible snapped, rescue would have been impossible. The record survived until August of 1949, when Otis Barton did a solo dive in a newly designed sphere, the benthoscope. He reached a depth of forty-five hundred feet and came back up to report his observations.[9] Unfortunately, despite all this work, none of these dives generated much new scientific data about deep-sea creatures.

In January 1960, Jacques Piccard, a submarine builder, and Lieutenant Don Walsh of the US Navy took four hours to sink 35,820 feet into the Mariana Trench. At nearly seven miles below the surface, they hit bottom and were stunned to see a bottom-dwelling flatfish. They stayed on the ocean floor for twenty minutes before coming back up again. In the last forty years, no one else has come close to these depths. It is a very expensive project, and there are no funding sources. The US Navy funded the 1960 expedition, and a Navy spokesman explained: "We didn't learn a hell of a lot from it, other than that we could do it. Why do it again?"[10]

The Navy did, however, provide some funding toward the *Alvin*, which is certainly the most productive deep-sea submersible in the world. It can reach depths of thirteen thousand to fourteen thousand feet and does about 150 dives a year. She is most probably best known for locating the *Titanic*. In 1966, two years after she was designed, the *Alvin* retrieved a hydrogen bomb that had been lost in the Mediterranean after a collision between a refueling plane and a B-52.

On another occasion, the steel cables connecting the *Alvin* to her mothership snapped in 1968, sending her five thousand feet down to the ocean floor. Fortunately, she was unoccupied at the time. It took eleven months before she was retrieved from the ocean floor off Cape Cod. Though there were no passengers on board, a bologna sandwich survived the ordeal. It was judged to be edible and was eaten after eleven months undersea.[11]

In 1982 *Alvin* found some very primitive bacteria living in deep-sea hydrothermal vents, where the conditions are very similar to those found on Earth long before photosynthetic organisms started producing an oxygen-containing atmosphere around the world.

Alvin has been so successful at discovering new deep-water species that there are several species that have a variation of the name *Alvin*. Though it is very expensive to do oceanography (renting the *Alvin* and her mothership, the *Atlantis*, costs $30,000 a day), the *Alvin* has been responsible for finding many new bioluminescent species and for vastly increasing our knowledge of life deep under the ocean surface. Yet we have better maps of Mars than we do of our own ocean floors.[12]

Just to give you an idea of how little we know about sea creatures, let us consider the marine worm *Xenoturbella bocki*. It lives fairly deep in the mud, a couple of hundred feet underwater. For more than five years, it was classified as a mollusk. Why? Because once or twice, a hungry worm ate some snails for lunch before it had the misfortune of being caught itself. Its DNA was analyzed and found to be that of a snail, and so in 1997 *Xenoturbella bocki* was classified as a snail. In 2003 Maximilian Telford of the University Museum of Zoology in Cambridge, United Kingdom, took a closer look at *Xenoturbella bocki* and realized that it had been classified on the basis of its lunch. The DNA that was sequenced was not that of the worm but that of its stomach contents.[13]

Bioluminescent bacteria are fairly common. Before antiseptic medications were available to disinfect wounds, bioluminescent bacteria would often be found in open wounds. This was a good sign because the bioluminescent bacteria are harmless and were displacing the disease-causing germs. In an 1899 thesis, a graduate student reported that he prepared a broth of bioluminescent bacteria and fed it to his cat, which survived. He thereupon drank some of the glowing broth himself and felt fine. Bioluminescent bacteria are also found in decaying matter, such as food and human bodies. It is quite possible that the luminescence from glowing bacteria on decaying corpses is responsible for many a ghost story. Aristotle was probably the first person to record this type of bacterial bioluminescence when he described the gleam on less-than-fresh meat and glowing dead fish on the beach.

There are only a handful of freshwater bioluminescent species, and there are no true glowing plants and higher vertebrates (besides fishes), such as frogs, reptiles, birds, and mammals. Scientists have used jellyfish and firefly proteins to make fluorescent flowers and have added some glow to certain porcine body parts, but they do not give off any light naturally, so they do not really count.

Lower terrestrial species that bioluminesce are fireflies, fungi, worms, and click beetles, among others. Since the vast majority of biotechnological and medicinal uses of bioluminescence are based on the light produced by fireflies and jellyfish, we will now devote ourselves to exploring interesting aspects of their bioluminescence.

Glowworms that emit light from their abdomens have fascinated and confused people for thousands of years. They are not really worms but are, in fact, fireflies or gnats in their larval forms. The "worms" that the inhabitants of New Zealand and Australia refer to as glowworms are actually larvae of fungal gnats. These interesting creatures live in tubular sheaths hanging from the roofs of caves and riverbanks. There, each larva spins thirty to forty silk threads that hang down from its home. Small insects attracted to the blue light emitted from the glowworm's tummy get caught in the silken threads, whereupon the glowworms, like a fisherman, reel in their threads and eat their meal. After the larval stage, which lasts ten months, the glowworms metamorphosize into adult fungus gnats that look very much like mosquitoes, except they produce a very faint glow.[14]

Fireflies are closely related to the glowworms or gnats of New Zealand and Australia. They are probably the most frequently observed bioluminescent organisms. This is because they are terrestrial, very common, and found on every continent except Antarctica. There are approximately two thousand species of fireflies in the world. More than 170 of these species are found in the United States, and most of them live east of Kansas. In 1974 Pennsylvania adopted *Poturis pensylvanica*, a firefly, as its official insect, even though no one has yet been able to prove that any *Poturis pensylvanica* spends its whole lifecycle in Pennsylvania.

Glowworms aren't really worms, and fireflies do not belong to the fly family. They are actually beetles, not flies. The main distinguishing characteristic of beetles that separates them from all other insects is the fact that they have hard wing covers. The name *beetle* comes from the Old English *bitan*, which means "to bite," and refers to the ability of beetles to chew and gnaw with their mandibles and maxillas. In contrast, true flies have spongy sucking-type mouthparts.[15] Firefly species are found all over the world, except in very cold or dry regions; none of them have spongy-sucking mouthparts, and they all have hard wing covers. Most species are nocturnal and spend the daylight hours in the shade of grasses, bushes, and trees. All juvenile flies glow, whereas there are some species in which the adult fireflies cannot glow. Every species of firefly also has a distinctive pattern of flashes. The intensity of the light produced largely depends on the species. It takes about six thousand females from the common European fruitfly (*Lampyris noctiluca*) to produce light of the same brightness as one candle, while only thirty-seven to forty females of the South American fruitfly (*Pyrophorus noctilucus*) are required to attain the same intensity.[16]

Incredibly, the fireflies' light emission is ten times as efficient as the light that is given off by a tungsten filament in a conventional lightbulb. Firefly bioluminescence is nearly 100 percent efficient, losing very little energy as heat during light production. This has resulted in some people calling the light emitted by fireflies "cold light." A lightbulb is only 10 percent efficient and loses a lot of its energy to heat, something I learned as a child when I used the heat generated by the lightbulb next to my bed to artificially increase the reading on the thermometer in order to appear feverish and miss some school.

Adult fireflies live no longer than two weeks, and they spend all their time and energy courting and mating. It is therefore not surprising that in most species their luminescence also has something to do with sex. When they are looking for a partner to copulate with, male fireflies fly around, flashing their light. Females sitting on the ground recognize the sequence of flashes as being characteristic of their species, a password of sorts. If they are interested, they will emit a sequence of flashes informing the male that they would like to meet. In many species, the males have large, protruding eyes so that they can readily detect even the faintest female response. The species-specific flashing sequence is an important evolutionary mechanism designed to prevent fireflies of different species from interbreeding and producing infertile offspring.[17] In the firefly world, brighter is better; researchers using artificial pulses of light have shown that female fireflies respond to males with the species-"correct" sequence and the brightest flashes.[18] Along with their sperm, males release a "nuptial gift," a high-protein nutritional emission that the females digest and pass along to their eggs as a care package. Researchers at Tufts University have shown that there is a correlation between the duration of the male's flashes and the size of the "nuptial gift" he provides.[19] Jim Lloyd, who took the photograph of the firefly shown in figure 2 in the photo insert, is a legend in firefly circles. Among many other interesting firefly discoveries, he found that fireflies that fly early in the evening give off a yellowish flash, while those that fly later emit green light. He correctly hypothesized that ecological pressures have forced the early flashing fireflies to develop a more yellow luminescence that is easily seen against the predominantly green background of the dimly lit foliage to attract mates.[20]

After the long-distance courtship has been completed and the couples have met, they do an "antenna dance." The purpose of this antenna touching is to confirm that the potential partner has the correct smell. If both partners are satisfied, mating starts and the lights are switched off. Some species of male fireflies spend no more than twenty minutes trawling the grass for a partner; others are more persistent and will spend several hours a night looking for a receptive mate. Despite all this effort, even the most successful males have no more than ten partners in their lives.

Fireflies have many natural enemies, such as birds, frogs, lizards, and spiders. The males of some species of fireflies have a noxious chemical that acts as a defense mechanism. The chemical structure of the material was found to be similar to that found in venomous poisons of Chinese toads and has been named "lucibufagin," a combination of the Latin words for *light* and *toad*. When the male firefly is challenged by a predator, it "reflexively bleeds"; the blood contains enough of the repellant lucibufagin to chase away the bad guys. In this way the males avoid being eaten.[21] Females of some species of fireflies do not have lucibufagin, and to add insult to injury, they cannot fly, making them even more susceptible to hungry predators. Females of the North American *Photuris* have found a rather nasty way to get over their vulnerability. Once they have found a mate of their own species and have successfully mated, they wait for a male from a different species to come flying by, looking for a mate. When the male sends out his interrogatory flash sequence, she is able to mimic the correct species-specific response for females from any one of eleven species.[22] The ill-fated male is unable to resist the false mating call and comes to try his luck in the antenna dance—a mistake, for she will devour him. In this way, female fireflies of the *Photuris* genus are able to obtain the lucibufagin. Unfortunately for the male fireflies, it does not affect the female *Photuris* firefly's digestion.

After mating, female fireflies lay their eggs in soft, damp soil. The eggs of some species glow continuously, they do not flash, and others emit no light. Firefly larvae hatch from the eggs. Some live on land, some in water; some glow, and some do not. The glowing larvae, as noted before, are often called *glowworms*. While the lives of adult fireflies revolve around sex, fireflies in the larval stage spend all their time and effort eating and growing. Most of the firefly's life is spent in its larval form, and it spends only a small fraction of its life as a pupa (the stage in which metamorphosis occurs) and then as an adult "fly."

Although there are many different species of fireflies, they all have the same chemical mechanism to produce light. Four chemicals have to come together in specialized cells found in the abdomen of the firefly before any light can be produced—oxygen, ATP, luciferin, and luciferase. The light comes from a small molecule called luciferin, which is made by the firefly inside its abdomen. However, for it to give off light, it needs to have some energy and to react with oxygen. The oxygen enters through tiny little holes in the abdomen of the firefly, and all the energy required for the light flash comes from ATP, an energy-rich molecule and an oxygenation reaction. Luciferase is an enzyme whose job it is to hold the luciferin, ATP, and oxygen in proximity to each other so that they can react. Once they have reacted and flashed their light, the luciferase is ready to take a new oxygen molecule, ATP, and luciferin and makes them react. In the firefly, this process is very controlled, and the light is given off in a precise Morse code–like sequence. If the poor firefly is crushed with a pestle and mortar, all four components will mix, and the resulting firefly mush will glow until one of the four components runs out. All of this occurs without any increase in heat; it is cold light.

The two most spectacular bioluminescent phenomena in nature must be the bioluminescent sea (as described in the introduction and shown in figure 1) and the synchronized firefly flashing observed in certain species of fireflies that are found in India, Malaysia, and the Philippines. In these areas, large numbers of fireflies collect on trees and flash their luminescent signals in near-perfect synchronicity, a natural Christmas tree with all its lights flashing on and off together. Though this phenomenon has fascinated scientists for hundreds of years, many did not believe published reports of the self-organized behavior. Scientists are still intrigued by the behavior and don't fully understand it.[23] Why and how do thousands of male fireflies flash their lights in perfect harmony?

In 1665 Christiaan Huygens, the inventor of the pendulum clock, while sick in bed watched two pendulums he had set up in his bedroom swing to and fro. He noticed that no matter how disjointed the swings of the two pendulums were initially, within half an hour, they were swinging in unison. No one could explain his observations until Michael Schatz, a physicist from Georgia Tech, used chaos theory to explain what was happening.[24] Chaos theory is a mathematical theory designed to find the underlying order in apparently random data, such as disjointed swings of the pendulums. The same chaos theory model derived by Schatz could possibly explain how spontaneous synchronization occurs in other areas, such as the synchronized firefly flashing described above and epilepsy resulting from synchronized nerve firing.

Let's go on to the jellyfish. There are over one thousand species of jellyfish living in the world's oceans. They are found in all the oceans, including the Arctic and Antarctic Oceans; some have even been observed at depths of more than thirty-two hundred feet below the surface, and many predate the dinosaurs. The adult jellyfish, with bell and tentacles, is called the *medusa*, which is named after the snake-haired goddess Medusa, whose looks turned people into stone. According to the *Guinness Book of Records*, the largest jellyfish is the Arctic Giant. One specimen that washed up in Massachusetts Bay had a medusa with a diameter of seven feet, six inches, and tentacles stretching 120 feet. The most venomous jellyfish is the beautiful but deadly Australian sea wasp. Victims die within three minutes of being stung by this jellyfish. Some jellyfish have evolved to become very efficient hunters. They don't go chasing after their prey; instead, they use their tentacles, which can be as long as sixty feet. These tentacles contain nematocysts, specially designed cells that shoot out a toxic-laden barb as soon as tiny hairlike triggers on the tentacles are touched.[25] Unsuspecting swimmers often inadvertently trigger the release of the barb and get stung by the jellyfish. But don't worry too much—the jellyfish do not go out of their way to sting humans and are not without their own enemies, especially the ocean sunfish, which can reach a weight of fifteen hundred pounds by eating little more than jellyfish. One thousand five hundred pounds—imagine the number of jellyfish eaten to get to that weight!

The jellyfish bell or umbrella-shaped bodies are composed of three layers: the outer exumbrella, the middle mesoglea, and the inner surface, which is called the *subumbrella*. The jellylike mesogleal material accounts for most of the body mass, hence the name *jellyfish*.

In the quiet waters underneath the waves, there is no need for a shell or skeleton. It makes much more sense to have a buoyant gelatinous body, particularly since the sticky gelatin and mucus is very useful for collecting food. If all the water were squeezed out of a jellyfish, you would find very little left, as a typical jellyfish is made up of 96 percent water, 3 percent protein, and 1 percent minerals. Jellyfish don't have any blood, bones, brain, central nervous system, eyes, ears, heart, lungs, gills, fins, head, tail, or teeth. To say that a jellyfish is heartless and brainless and that its mouth is no different from its anus is not an insult to a jellyfish but just the plain, unvarnished truth.

When there is lots of food around, jellyfish grow rapidly larger; unfortunately, this phenomenon has also been observed in humans. However, when there is a food shortage, jellyfish have the unusual ability of shrinking so that they need less food to survive. Talking about food reminds me of the upside-down jellyfish, *Cassiopeia xamachana*, which doesn't have a central mouth. It lives in shallow water, where it turns itself upside-down and uses suction to stick to the ground with its fused oral arms pointing upward looking for prey. On the oral arms are hundreds of tiny mouth openings connected by channels to the stomach.[26]

If all jellies are brainless, you might want to know how all the jellyfish that swim the "right" way up manage to do it. They have balancing organs all around the edge of their umbrellas. These contain very small, stonelike particles. When the jellyfish tilts, the granules touch cilia (little hairs) that line the "balancing organs," which stimulates the jellyfish to adopt a more horizontal position.[27]

As interesting as some jellyfish are, it is the less prepossessing *Aequorea victoria* species of jellyfish that is largely responsible for the glowing genes revolution. It is the source of a glowing protein called *green fluorescent protein* (GFP). *Aequorea victoria*, also known as the crystal jelly, is commonly found off the West Coast of North America, from central California to Vancouver, mainly near Washington and British Columbia. The medusa of *Aequorea victoria* can grow up to ten inches in diameter, but it is usually no larger than four inches in the Puget Sound, where most of the initial green fluorescent protein research was conducted. Unfortunately, it is impossible to spawn a revolution and not be the center of at least one controversy, and so it is with *Aequorea victoria*, which has had a bit of an identity crisis. For nearly a century, a healthy debate has been raging about whether *Aequorea victoria* is a separate species or whether it is just a variant of *Aequorea forskalea*, also known as *Aequorea aequorea*. In this book, I will use the names *Aequorea aequorea* and *Aequorea victoria* interchangeably.

Jellyfish that are found in deeper waters are more likely to bioluminesce than species that are found closer to the surface. Most jellyfish bioluminescence is used as a defense against predators. "Comb" jellyfish produce bright flashes to startle a predator; others can produce a chain of light or release thousands of glowing particles into the water to confuse the predator. There are even some jellyfish that can discard their tentacles to act as glowing decoys. Others make a glowing slime that can stick to a potential predator and make it vulnerable to its predators. Scientists are not certain why *Aequorea victoria* fluoresces. It doesn't seem to be for the important things in life—reproduction, food, or self-preservation.

CHAPTER 2

FROM PLINY'S WALKING STICK TO BURNING ANGELS

Before looking at how the genes from glowing organisms have started a biotechnological revolution, let's go back in time and see when light-producing organisms were first described in literature. Pliny the Elder (23–79 CE) would be a good person to start with. If he were alive today, perhaps he would have a television show on the Food Network, for he found and devoured an edible clam that squirted phosphorescent green slime when it was frightened. Anyone who ate the clams ended up with a pair of glowing green lips. The Romans got a kick out of this bizarre spectacle, and for a while, glow-in-the-dark banquets were a popular first-century eating craze. Unfortunately, there is no record of how these clams were prepared. Were they eaten raw, soaked in red wine, or were they served as part of a luminescent clam chowder? Pliny also reported that a paste made from the luminous materials of the clam mixed with flour, honey, and water would produce light, even up to a full year, as long as water was added. The clams were so popular that they are nearly extinct today. In book nine of his *Historia Naturalis*, Pliny writes: "They shine of themselves in the dark night, when all other light is taken away. The more moisture they have within them, the more light they give:

insomuch as they shine in men's mouths as they be chewing of them: they shine in their hands: upon the floor, on their garments, if any drops of their fatty liquor chance to fall: so it appears."

Aequorea victoria need not be concerned that they will be driven to extinction due to an upcoming twenty-first-century glow-in-the-dark jellyfish-eating craze. Their bioluminescent organs are located on the very edge of their umbrella. One would have to eat a bucket full of *Aequorea victoria* to get a fluorescent mouth and face, and so it would not be nearly as amusing to eat luminescent jellyfish as it is to eat clams that have glowing slime. However, jellyfish are considered a delicacy in Asia, and there are some plans to try to introduce dried Cannonball jellyfish into the American food market. Dried jellyfish are nice and crunchy and rich in minerals. For thousands of years, Asians have consumed dried jellyfish and have used jellyfish to treat high blood pressure, arthritis, and bronchitis. Today, Japan and South Korea are still the two largest jellyfish consumers.[1] Perhaps the Florida leatherback sea turtles and sunfish will soon have some competition for their jellyfish diet.

Pliny the Elder saw many parts of the world during his military career, giving him the opportunity to observe many new and exotic species. His descriptions of light-producing organisms were among the most complete and accurate for his time. Before he died in the eruption of Mount Vesuvius, he described the luminescence of snails, jellyfish, lantern fish, and fungi. He reported that glowworms "shine in the night like a sparkle of fire" and that luminous fungi could be "very effectual in medicine and antidotes." He continued: "It grows on the head and tops of trees: it shines by night, and by the light it gives off in the dark, men know where and how to gather it."[2]

Pliny was the first person to record the luminescence of the glowworm and the bioluminescence of jellyfish. He described the light of *pulmo marinus* (sea lung),* a purple jellyfish. He was also the first to find practical applications for the light produced by some organisms. Not only did Pliny describe using the luminescent slime of clams to amuse fellow party goers, but also he was the first person to report a use for jellyfish bioluminescence: "Rub a piece of wood with the fish called Pulmo Marinus, it will seeme as though it were on fire; in so much as a staffe so

---

* Because of their pumping-swimming action, the jellyfish were called *pulmo marinus*, or "sea lungs," by the Romans.

rubbed or besmeared with it, may serve instead of a torch to give light before one."[3] Little did he know that the very same bioluminescence of Pulmo Marinus's cousin, *Aequorea victoria*, would light the way for scientific discovery in the twenty-first century.

As long people walked the Earth, I am certain that they have seen and been intrigued by light emitted by decaying food, luminous rocks, and light-producing organisms. Pliny the Elder certainly was not the first person to describe organisms that produce light.

The first reports of bioluminescence are in ancient Chinese and Indian literature. The line "I-yao hsiao-hsing" in the *Shih Ching* from 1500–1000 BCE has been translated as both "The fitful light of the glowworms would be all about" and "Glowing intermittently are the fireflies."[4]

Interestingly enough there is no record of any luminescence in the Bible, Talmud, or Koran. Early Indian and Arabic literature have numerous references to the firefly, but they were not very impressed by the amount of light it produced. In fact, the early Arabic name for a firefly is derived from the word for a man who is so stingy that he makes a fire too small to be of any use. There are no known Egyptian hieroglyphic characters for glowworms, fireflies, or luminescence.

Four hundred years before Pliny was buried in the ashes of Mount Vesuvius, Aristotle first made mention of the fact that luminescence was "cold light." By this he meant that the light produced by some jellyfish, fireflies, and fungi is not associated with an increase in heat. Luminescence is still referred to as cold light today.

Since Pliny and Aristotle there have been many well- and lesser-known scientists who have investigated luminescence. Here we will provide just a taste of the history of luminescence and focus on the most interesting observations and the contributions of some of the most famous researchers. Then, we will focus on the light emission of fireflies and jellyfish.

Alchemists in the sixteenth and seventeenth centuries strived to find a way of converting ignoble metals into gold. As part of his quest to find a philosopher's stone that turned other elements to gold, Vincenzo Casciarolo, a cobbler and part-time alchemist, found a stone in 1602 "which would not lend itself to produce gold, but which would absorb the golden light of the sun, like a new Prometheus stealing a Celestial treasure."[5] It was impossible to change the stone into gold, but it absorbed light and would reemit it in darkness. The stone was found on the outskirts of Bologna and was named the Bolognian Stone. Its phosphorescence intrigued many of the great minds of the time. Galileo Galilei was one of them, and his explanation for the light emitted by the Bolognian stone was that it must have been exposed to some heat and light, which it had absorbed like a sponge and was now releasing.[6]

Earlier we briefly mentioned the "world's brightest glowing bay," Vieques, Puerto Rico. Glowing seawater has been observed throughout recorded history; in 500 BCE, Anaximenes, a Greek philosopher, was one of the first to report sea phosphorescence. He believed that lightning and the light produced when one strikes the ocean with an oar were very closely related. Aristotle in his *Meteorologia* (book 2, sec. 9) reports that Cleidemus, an early historian, had similar opinions, saying that "lightning is nothing objective but merely an appearance. They compare it to what happens when you strike the sea with a rod by night and the water is seen to shine. They say that the moisture in the cloud is beaten about in the same way, and that lightening is the appearance of brightness that ensues."[7]

About two thousand years later, Benjamin Franklin demonstrated an important characteristic of a good scientist—the ability to change one's opinion if the evidence so requires. In 1747 he publicly supported the view that one could consider "the sea as the grand source of lightening, imagining its luminous appearance to be owing to electric fire, produced by friction between particles of water and those of salt."[8] Franklin knew a great deal about lightning; after all, he foolishly flew his kite in a thunderstorm. Because of his expertise, Franklin's support of the electrical nature of the phosphorescence of the sea held sway. But, unfortunately, he lived very far from the ocean and had no opportunity to do experiments with seawater. By 1753 he had changed his opinion on the origin of the luminous appearance that is occasionally observed in seawater. In his book *Experiments and Observations on Electricity Made at Philadelphia in America, to which Are Added Letters and Papers on Philosophical Subjects*, he published a letter from James Bowdoin, the governor of Massachusetts, in which he reported that filtering phosphorescent seawater through a cloth resulted in the removal of the light from the water and that the "said appearance might be caused by a great number of little animals, floating on the surface of the sea, which on being disturbed, might, by expanding their fins, or otherwise moving themselves, expose such a part of their bodies as exhibits a luminous appearance, somewhat in the manner of a glow-worm, or firefly."[9] Benjamin Franklin withdrew his support for the electrical theory and fully supported the luminescent organism theory, which is indeed the correct explanation.

In Vieques, Puerto Rico, a gallon of water can contain as many as 720,000 microorganisms called *dinoflagellates* that flash a greenish-blue light when agitated. The light is presumably a defense mechanism that these tiny sightless "water fireflies" have developed to dazzle their enemies. The island of Vieques is twenty-two miles long, and its claim to fame has been the dispute over the US Navy's using part of the island as a firing range for its ships and planes. It stopped this practice in 2002 after inadvertently killing an innocent security guard. Luckily, the bioluminescent bay is on the far side of the island and has been unharmed by fifty years of bombing that only recently ceased. The 160-acre bay is the last glowing bay left in the Caribbean. Unfortunately, due to pollution all over the world, the number of bioluminescent organisms that produce glowing seawater has been dramatically reduced everywhere, and Vieques has become a unique and endangered site.

In California, Michael Latz, a marine biologist at the Scripps Institution of Oceanography, had heard many stories of dolphins swimming through ocean water that was full of dinoflagellates, leaving beautiful luminescent trails. Being a practical soul, he immediately thought that this would be a great way of analyzing the dynamics of the dolphin's swimming motion. It was a very romantic idea, but not very practical, since it wasn't easy to find dolphins and high enough concentrations of dinoflagellates at the same time and place. But Latz did not give up, and he went to the US Navy research facility at San Diego Bay, where there are dolphins trained to save people and retrieve mines. The dolphins are kept in large pens and have been conditioned not to fear people. Michael Latz and his students put covers over the pens so that they could observe all the dinoflagellant bioluminescence without having to require high concentrations of dinoflagellates. The bioluminescent layer they observed around the dolphins was surprisingly thin; evidently dolphins are able to cut through the water creating very little turbulence. Even more amazing was the finding that there was an area around the dolphin's nose that was curiously free of bioluminescence. Latz suggests that there is little water movement around this area because this is where the dolphins sonar location system is located.[10]

As exciting as these new insights and general bioluminescence are, let us return to the past and now focus the rich history of the lowly firefly. In 400 BCE, the firefly was called *Ying-lao* and *Chi-chao*, names still used today. It was recorded that fireflies appear in the third month of summer and that they arise from decaying grasses—a belief that survived in China for at least two thousand years. The same concept has also appeared in Japanese literature where a 1712 encyclopedia classified different species of fireflies by the decaying grasses and woods from which they arise. In Japanese poetry, the firefly has also been a metaphor for passionate love and hearts on fire since the eighth century. Hundreds of haiku have been written about the "burning angels." In Japanese mythology, the firefly's eerie lights are also thought to symbolize the souls of the dead. Firefly viewing was most popular in Japan during the Edo Period (1603–1867), but still today many Japanese people go to watch fireflies, and there is an annual firefly festival outside of Kyoto.

Fireflies can also be found in modern poetry. One of my favorites is the following limerick by written by Ogden Nash in 1937; not surprisingly it is called "The Firefly."

> The firefly's flame
> Is something for which science has no name.
> I can think of nothing eerier
> Than flying around with an unidentified glow on a person's posterior.

During his circumnavigation of the Earth, Sir Francis Drake (1540–1596) observed tropical fireflies alighting on some trees: "Amongst these trees, night by night, through the whole land, did show themselves an infinite swarm of fiery worms flying in the air, whose bodies being no bigger than our common English flies, made such a show and light, as if every twig or tree had been a burning candle."[11]

Francis Bacon (1561–1626) went to Trinity College, Cambridge University, at the tender age of twelve. He had an up-and-down sort of life; his father, the Lord Keeper of the Seal of Elizabeth I, died when Francis was eighteen, leaving him penniless. He recovered to become Lord Chancellor of England, only to be accused of accepting a bribe, for which he was found guilty and lost both his money and position. During his life, he devoted some time to doing scientific research, and he summed up what was known in the 1620s about the firefly:

> The nature of the glow-worm is hitherto not well observed. Thus much we see: that they breed chiefly in the hottest months of summer; and that they breed not in fields, but in bushes and hedges. Whereby it may be conceived, that the spirit of them is very fine, and not to be refined by summer heats: and again, that by reason of the fitness, it doth easily exhale. In Italy, and the hotter countries there is a fly they call Lucciole, that shineth as the glowworm doth; and it may be it is the flying glow-worm. But that fly is chiefly upon fens and marches. But yet the two former observations hold; for they are not seen but in the heat of summer; and sedge, or other green of the fens, give as good shade as bushes. It may be the glow-worms of the cold countries ripen not far as to be winged.[12]

Francis Bacon died in 1626 from bronchitis, following a cold caught while stuffing a dead chicken full of snow to see whether the colder temperature would slow its rate of decay. His supposition that the fireflies found in the warmer Mediterranean countries are related to the glow-worms found in England was correct—glowworms are the larval forms of fireflies. In some cases the larvae glow; in others, they do not.

Fireflies may have sparked many romances, played a role in literature, and possibly even changed the course of history. In 1634 the English navy was sailing past Cuba, where it wanted to land, but it saw many flashes of lights coming from the shore. Thinking that the Spanish forces had beaten the navy to shore and were lighting campfires, the navy decided to sail on. History shows us that there were no Spanish forces in Cuba at the time. The Brits must have mistaken firefly flashes for a large military force spread across the island.[13]

Before we discuss the jellyfish, let's look at some of the early chemical studies of firefly luminescence. Robert Boyle (1627–1691) was the originator of Boyle's Law, which states that the volume of a gas varies inversely with pressure. He was also the first to realize that there is something in the air that was required for firefly luminescence, since no light is produced in a vacuum.

Another chemist known to most science students is Michael Faraday (1791–1867), perhaps the best experimentalist who ever lived. Faraday had very little formal education; after elementary school, he went to work as an apprentice for a bookbinder. There he became interested in chemistry and physics. After attending a talk given by Humphry Davy, he managed to convince Davy to take him as an assistant. Davy was a great chemist responsible for discovering twelve new elements, including calcium, potassium, and sodium. He also discovered aluminum, which is pronounced differently in England and America. When he discovered the element in 1808, he called it *alumium*, a name he never liked much. So four years later, Davy decided to change it to *aluminum*, which the Americans have used ever since. The Brits weren't happy with either name and call the element *aluminium*.[14] Davy was very interested in the physical properties of light, but he never studied luminescent organisms. He left that to his protégé, Michael Faraday, who, in 1814, reported that the glow of crushed glowworms lasted for several days and was therefore not related to the glowworm's life but to some substance inside it. He continued that the luminous matter

> is yellowish-white, soft and gelatinous. It is insoluble in water and alcohol. It does not immediately lose its power of shining in them, but it is sooner extinct in alcohol than in water. Heat forces out a bright glow, and then it becomes extinct; but if not carried too far, the addition of moisture after a time revives its power. No motion or mixture seems to destroy its power whilst it remains fresh and moist, but yet a portion thus rubbed, sooner lost its light than a portion left untouched.[15]

Not long after Faraday described the luminous material in fireflies, his mentor Davy died. For many years, he had been addicted to inhaling nitrous oxide, and in 1829 he died an early death probably brought on by the excessive use of laughing gas.[16]

This sampling should provide an idea of how fireflies were viewed in the past. Let's go on to see how jellyfish luminescence was treated. It took a long time before anyone got around to improving and expanding on Pliny the Elder's descriptions of jellyfish luminescence. In 1646 Athanasius Kircher (1602–1680), a German Jesuit priest, wrote *Ars Magna Lucis et Umbrae*, which Joseph Priestley, the discoverer of oxygen, called "a very capital performance." Kircher was one of the major thinkers of his time; he taught ethics, mathematics, Syrian, and Hebrew at the University of Würzburg. Like Pliny he was fascinated by Mount Vesuvius. He wanted a close-up view of the volcano, so he climbed up the mountain and let himself down into the crater to measure its dimensions. Unlike Pliny, Kircher survived his Vesuvius experience. He wrote more than fifty books in his life. Although I have no doubt that he was an excellent naturalist, I believe he was a little harsh on the jellyfish:

> Mention should be made here of another marvel of the sea, which, although it is nearly the lowliest and most despised of blood-containing-animals, yet has not a little nobility by virtue of its innate light. Some call it the *Pulmo marinus* [sea lung]. . . . But it is a marvel that the liquid of this *Pulmo* when rubbed on black sticks and certain other things causes them to shine in darkness no differently than fire: I discovered this by experiment. . . . I think that this is the reason why nature wanted to imbue these animals with light: namely, that they should not live in perpetual darkness and seem to have been provided with eyes by nature in vain, since they live in the depths of the sea and cling to sticks, but the depths of the sea are dark and are not reached by rays of the sun, as divers inform us. Thus, nature gave to these animals viscous liquid imbued with counterfeit light, that by its help, as it were by a lamp born with them, they might seek food and also easily elude the snares of foes by the voluntary emission of light and darkness, and thus they might not be destitute of these things that are necessary for their own life.[17]

A little more than a century after Kircher's description of the sea lung's luminescence, another German, the scientist and explorer Alexander von Humboldt (1769–1859), undertook the first recorded jellyfish luminescence experiments:

> On the morning of the 13th of June, in 34 degrees 33' latitude, we saw large masses of this last mollusca in its passage, the sea being perfectly calm. We observed during the night that, of three species of medusas which we collected, none yielded any light but at the moment of a very slight shock. . . . If we place a very irritable medusa on a pewter plate, and strike against the plate with any sort of metal, the slight vibrations of the plate are sufficient to make this animal emit light. The fingers with

which we touch it remain luminous for two or three minutes, as is observed in breaking the shell of pholades. If we rub wood with the body of a medusa, and the part rubbed ceases shining, the phosphorescence returns if we pass a dry hand over the wood. When the light is extinguished a second time, it can no longer be reproduced, though the place rubbed be still humid and viscous. In what manner ought we to consider the effect of the friction, or that of the shock? This is a question of difficult solution. Is it a slight augmentation of temperature, which favors the phosphorescence, or does the light return, because the surface is renewed, by putting the animal parts proper to disengage the phosphoric hydrogen in contact with the oxygen of the atmospheric air.[18]

Another early experimentalist worth recalling is Abbe Lazzaro Spallanzani (1729–1799), who included an article on *Medusae fosforische* in his *Viaggi alle due Sicilie* (1793).[19] He found that the material responsible for luminescence in jellyfish was a mucus formed on the edge of the umbrella and on the arms of the jellyfish. When the mucous was removed from the jellyfish, it could not be made to give off light in any way. Spallanzani wrote, "Another medusa, which was dead, and had not been luminous for some time, was lying, out of the water, in the window of my chamber during the night. A slight rain chanced to fall, and every drop which fell on the dead medusa was changed into a brilliant spangle, till in a short time the medusa was studded all over with such shining points. I could produce no such effect by sprinkling it with sea water in imitation of rain." He goes on to note that the luminescence of the dead medusae in milk was so great "that I could read the writing of a letter at three feet distance."[20]

It should be noted that Spallanzani did more than jellyfish experiments; he was the first person to realize that digestion of food in the stomach is more than just mechanical grinding—and that it is a chemical process. In 1780 Spallanzani performed artificial insemination experiments on dogs, frogs, and silk moths. I am sure that he would have been fascinated to know that 220 years later, jellyfish genes would be inserted into a monkey to produce, by artificial insemination, a monkey that has glowing toenails.

These were some examples of the first experiments with bioluminescent materials, fireflies, and jellyfish. In the sixteenth and seventeenth centuries, luminescent materials were added to medicines to impress the patients. Phosphorus was the most common additive. Alphonse Leroy was a French doctor so fond of using phosphorus in his treatments that an autopsy of one of his patients revealed that he had luminous internal organs. Leroy practiced what he preached. He once took three grams of phosphorus in a syrup himself. At first he did not feel well, which is not surprising, since phosphorus is a poison, but on the next day, he reportedly felt that his strength was doubled and his sexual appetite greatly improved.[21] These observations led to phosphorus being prescribed for impotence by many other physicians. For example, in *Zoonomia* (1794), Erasmus Darwin (1731–1802) reported that phosphorus could be used "to restore and revive young people exhausted by excesses." Since it is a poison, however, I would not suggest that you try this at home.

Now that we have seen how people have thought about bioluminescence throughout history, let's explore how researchers discovered the way in which the firefly emits light and how this knowledge has led to innovations in areas as disparate as the food industry and space research.

CHAPTER 3

USING FIREFLIES TO LOOK FOR LIFE ON MARS?

Fireflies crashing into William David McElroy's face might have been responsible for the fact that firefly proteins may be traveling up to Mars one day and that they might be used to look for life on the red planet.

Bill McElroy, born in 1917 in Rogers, Texas, was raised on a farm and walked barefoot to a one-room school. In high school, he excelled at football and went to Stanford on a football scholarship, where he played right end on the 1938–39 team.[1] McElroy wanted to study medicine, but a genetics course with the Nobel Prize winner George Beadle turned him toward basic research instead.[2] According to his obituary, Bill first became interested in luminescent organisms when fireflies attracted by the glow of his cigar began crashing into his face.[3] Why did they waste so much energy-producing light, McElroy wanted to know, especially since luminescence seemed fairly unusual in nature. In 1941 E. Newton Harvey at Princeton University was the undisputed world expert on bioluminescence, so Bill McElroy went to Princeton to get his doctorate under Harvey's tutelage to learn more about firefly light emission. It was the start of a forty-plus-year career devoted to studying the luminescence of living species. From 1946 to 1969, armed with his PhD, he went to

Johns Hopkins University, where he soon became the department chair. While at Hopkins, McElroy started his own firefly research group. Like most early firefly bioluminescence researchers, as well as scientists in many fields, McElroy was interested in finding the answers to some interesting basic scientific questions that did not have any obvious practical applications. His main preoccupation was in finding how the chemical reactions that occurred in the firefly produced flashes of light. It had been known for some time that at least two compounds were required to produce the firefly's flashes of light. But McElroy knew that there had to be more to it than that.

The French physiologist Raphael Dubois had done most of the early firefly research. He was the one who discovered that firefly luminescence required the reacting of at least two different chemicals. In 1885 he ground up some firefly abdomens and found out that by adding cold water to the firefly mush, he could make a glowing solution that started off shining brightly but became fainter with time. If he added hot water to the crushed abdomen puree, it did not glow. However, if he waited for the hot solution to cool, and then added some of it to the spent solution that was no longer glowing, it was reinvigorated and started to glow again. Dubois hypothesized that both solutions contained two important components. In the cold solution, they were both intact and gave off light until one of the two components ran out. In the hot solution, the heat destroyed one of the components, and no light was produced. When it was cooled and added to the spent cold solution, that solution started glowing because the component that had survived the heat was the one that had been used up in the cold solution. He named the molecule that was consumed in the light-producing reaction *luciferin* and the component that was destroyed in the hot water *luciferase*. They were both named after Lucifer, the fallen angel of light.

E. Newton Harvey, McElroy's doctoral adviser, conducted research based on Dubois's studies. Newton Harvey was the first modern scientist to approach bioluminescence from a chemical point of view. He was born in Germantown, Pennsylvania, and was fascinated by nature from a very early age, collecting every conceivable natural object, including frogs in the family bathtub to lay eggs in the spring. He devoted four books and more than 125 published papers to bioluminescence. His early work laid the foundations of the glowing gene revolution. Newton Harvey found that most bioluminescent organisms require at least two components for light production—a luciferin that is consumed in bioluminescence and a luciferase that is heat sensitive. However, the chemical structures of the luciferins are not the same in all species.

In 1946, when McElroy started his laboratory in Johns Hopkins, many questions about firefly luminescence remained unanswered. Were luciferin and luciferase the only two molecules required for light production? What determined the brightness of the light flashes? McElroy was the man to answer these questions. To find out what was happening inside the fireflies to create their distinctive flashes of light, McElroy needed lots of fireflies, many more than he and his students could possibly hope to catch themselves. McElroy was clearly a resourceful man—he decided to recruit local school children to do his firefly hunting for him. In 1947 he offered a bounty of twenty-five cents for every hundred fireflies that were brought to his labs in Mergenthaler Hall, on the Johns Hopkins Homewood Campus, as well as a ten-dollar prize for the person who handed in the most fireflies.[4] In the first year of the scheme, McElroy netted forty thousand fireflies, and ten-year-old Morgan Bucher Jr. won the ten dollars.

The light-producing reactions in the fireflies occur in special cells called *photocytes*, which are located in the fireflies' abdomens. The photocytes, or light cells, are layered between two sets of cells, a thin, nearly transparent outer layer that allows the light produced in the photocytes to pass out of the firefly unhindered and an inner layer filled with uric acid crystals that reflects all the light produced by the light cells out of the abdomen.[5] Laboratory assistants and graduate students had to dry thousands of fireflies and then separate their abdomens, where the photocytes were located, from their heads.

McElroy's army of firefly catchers steadily grew in size, and by the 1960s, he was collecting between five hundred thousand and a million fireflies a year. The *Baltimore Sun* reported, "Life in Baltimore, for fireflies, became a chancy, precious thing."[6] However, there was no cause for concern, unless you were a male firefly, that is, for in 1965, McElroy told a *Baltimore Sun* reporter, "Some people have complained that our collections might cut deeply into the firefly population, but the flies we are collecting are males. The females stay in the grass and lay eggs; so our collection should have no effect on the population."[7]

J. Woodland "Woody" Hastings, a professor at Harvard University, is one of the most well-known bioluminescence researchers active in the field today. He did his doctoral work with E. Newton Harvey and his post-doctoral studies with McElroy at Johns Hopkins. One of his first jobs in the lab was supervising McElroy's firefly collecting operation. His task was to pay the collectors for their flies and put the fireflies in a vacuum desiccator to dry them out. He started working for McElroy in 1951 and has written an amusing and easy-to-read article about his work with McElroy, titled "Firefly Flashes and Royal Flushes: Life in a Full House."[8] The title refers to the fact that McElroy was a dedicated poker player. Several evenings a week, the dean of students, a professor of engineering, the editor of the Hopkins magazine, McElroy, and others met in the firefly lab to play poker. They chose to play in the firefly lab because it had one of the first window air conditioner systems at Johns Hopkins University. The air conditioner, built and designed by Hastings, had been bought to cool a high-voltage photomultiplier, not the poker players, but they made good use of it.

In his job as firefly broker, McElroy trained Hastings in bargaining with twelve- to fifteen-year-old firefly collectors. They would come to Hopkins with hundreds of fireflies and demand payment for ridiculously inflated numbers of fireflies. They had so many flies that counting was difficult. Weighing didn't work because the dishonorable collectors added twigs and stones to their fireflies; some even wet them. So McElroy, the poker player, had devised a new strategy, and it worked something like this:

"How many fireflies?"

"Seven hundred."

"Ridiculous, more like three hundred."

"Aw, come on at least six hundred."

"All right, I'll pay you for five hundred."

"But there are more than five hundred."

"OK. Double or nothing. I'll pay you for five hundred. If you want to count them and there are more than five hundred, you get paid for one thousand. If there are less than five hundred, you won't get paid at all."[9]

McElroy instructed his buyers in this negotiation technique with instructions to always overestimate the number of fireflies, so that the firefly lab would never lose. It worked and streamlined the bargaining process.

In 1961 chemists at Johns Hopkins University managed to synthesize luciferin for the first time, and soon Sigma Aldrich, a large St. Louis–based chemical company, started to sell both luciferin and luciferase. Although the luciferin was synthesized in the lab, fireflies were still required to obtain the firefly luciferase. So Sigma, much like McElroy, started a firefly club that paid kids to collect fireflies in order to get as many firefly abdomens as possible.

Today, many vendors sell luciferin and luciferase, but fireflies are no longer being pureed to obtain the molecules. Instead, new techniques (described later) are being used to trick bacteria into making luciferase, and then scientists harvest it from the bacteria. One of the people responsible for this recipe was Marlene DeLuca, Bill McElroy's wife.

In 1961, the same year that luciferin was made in the laboratory for the first time, Bill McElroy took on a new postdoctoral assistant named Marlene DeLuca. She was very successful in the laboratory, and after her postdoc work, she briefly was employed as an assistant professor of biology at Johns Hopkins, but she resigned when she and Bill McElroy got married. She moved to Georgetown University, where she continued studying the chemical reactions luciferase underwent in order to produce flashes of light.[10] Marlene was not only an important part of McElroy's life, she was also a major contributor to the scientific research that resulted in luciferase being used as a probe in biological systems. When McElroy moved to the University of California, San Diego, to become the chancellor, Marlene moved with him and became a member of the chemistry department. One day, Keith Wood, one of her doctoral students, came to her and asked her if he could clone the luciferase gene. Marlene was not impressed with the idea and reluctantly let Wood do the project, but she gave him only six months to do it. Despite the time constraint, he succeeded. This convinced Bill McElroy and Marlene DeLuca that this was a worthy research direction.

Of the more than a thousand firefly species that could be cloned, one of them, found only in Jamaica, is the only firefly that can emit two different colors. DeLuca and McElroy flew to Jamaica to catch some of these unique fireflies so that Keith Wood could clone them. When they got to the Caribbean island, they found that they had come during the wrong time of the year—there were no fireflies. Fortunately, DeLuca and McElroy contracted a Jamaican inhabitant to catch the fireflies and send them to San Diego.

Besides cloning luciferases from a variety of firefly species, Wood and DeLuca also managed to insert the luciferase gene into *Escherichia coli*, intestinal bacteria, which were then able to make firefly luciferase.[11] They also placed the luciferase gene in some plants, which would glow when watered with a solution containing luciferin. Wood, DeLuca, and the glowing plant were mentioned both in *Time* and on the *Tonight Show*.

What did Bill McElroy contribute to glowing-gene technology? When McElroy started his firefly research in the mid-forties, the time was right for making an important finding. In 1941 Fritz Albert Lipmann and Herman Kalckar had just discovered that a molecule called adenosine triphosphate (ATP) was an extremely important source of energy in all living organisms. McElroy wondered whether ATP could be the energy source for the light produced by fireflies. It turned out that it was! He found that when he added ATP to mixtures of luciferin and luciferase that had been obtained from firefly abdomens, the mixtures started glowing. The more ATP he added, the brighter the firefly extract glowed. McElroy had thus found another component that was required to produce light in fireflies—ATP. He was lucky, for ATP is not required in all bioluminescent systems.

ATP is a molecule that provides the energy for heat, nerve conduction, and muscle contraction. It is responsible for making our hearts beat and our brains think, so it should not be surprising that it can also provide the energy to make firefly tails light up. Although ATP can supply the energy that is converted into light in the firefly, McElroy had not yet shown that there was ATP in their light cells and that ATP was being used by living fireflies to flash their amorous signals. In order to show that ATP was indeed the source of energy, McElroy and his graduate students set out to isolate and purify the active ingredients of firefly light cells. It took thousands of fireflies and thousands of hours of tedious and painstaking work before McElroy could show that ATP, oxygen, luciferin, and luciferase were required for producing flashes and that they were all present in the light cells. It was worth all the effort though because now the firefly luciferin/luciferase system could be used to detect ATP. This may not sound very exciting, but I assure you that this was a major breakthrough.

All known forms of life on Earth use ATP as a source of energy. It is an instant source of energy and is not used to store energy. Carbohydrates and fats are used for long-term energy storage and are converted to ATP when the energy is needed. It's a bit like electricity—we have all kinds of power plants that can convert long-term energy forms like coal to instant energy sources like electricity, which then comes out of our sockets to power our computers, fridges, and so forth.

ATP is an amazing molecule; right now every cell in your body will have about two billion ATP molecules in it. As you need energy, the ATP molecules will be broken down. Within two minutes, they can all be used up and replaced by new ATP molecules. In a typical day, you can produce and use up half your body weight in ATP.[12] ATP is a universal power socket; if something is alive, it has ATP. If it has ATP, the addition of luciferin and luciferase will produce light in the presence of oxygen. We now have a test for life.

In the 1960s, luciferase systems were used as indicators of life for the first time. They were not used by morticians to make sure that their bodies were indeed ready for embalming—instead, they were used to see whether there were any microbial life in areas that were supposed to be clean and sterile. At the time, luminometers—the instruments required to measure the amount of light given off by the luciferase system—were very large and expensive, and ATP bioluminescence techniques for hygiene monitoring were barely used. In early versions of the luciferase-based microbe test systems, the luciferin and luciferase were supplied as freeze-dried powders, which were dissolved in water to make enough solution for twenty-five to fifty tests. It was an expensive and cumbersome procedure and not used much.

Today, bioluminescence tests for microbes are routinely used. They are much quicker than the alternative test, which is to take a swab from the location of interest and grow a cell culture. The luciferase/luciferin system is still provided as a freeze-dried powder, but it now has a much longer shelf life. The United Kingdom Defense Science and Technology Laboratory has biochemically modified firefly luciferase so that it can go through numerous freeze-drying cycles and is stable in the liquid form for months, even at room temperature. Biotrace, a British company, has the exclusive license for these modified firefly enzymes and uses them in its Lucigen range of ATP detectors.

Total ATP bioluminescence tests can be done in less than a minute and are a vital method to measure the effectiveness of disinfection. If a surface has been completely disinfected, it will have no live bacteria or virsuses on it, and luciferase will not glow. Total ATP bioluminescence tests have been employed in detecting microorganisms in a wide range of samples, from beer to urine. Many companies, like the Massachusetts-based Imsco, which has a contract to test beverage ingredients of Coca-Cola, Pepsi, and 7 Up for yeast, mold, and bacteria, use luciferase-based tests.

Kasthuri Venkateswaran, a Caltech researcher, has published a study that contrasted the effectiveness of cultivating bacterial colonies and doing colony counts with modern firefly luciferase bioluminescence ATP assay methods. These methods differentiate between extracellular ATP (dead cells) and intracellular ATP (live microorganisms). He found that contamination by living microorganisms could be reliably measured by the luminescence of firefly luciferase and that it was much faster than using the conventional method, which is based on cultivating the microorganisms and then counting the colonies.[13] Commercially available bioluminescent ATP-detecting devices can register as little as 1×10^{-12} grams of ATP. This is equivalent to a thousand bacterial cells. Some detectors used in research are even more sensitive.

Is there life on Mars? The European Space Agency (ESA) and NASA have just sent a number of missions to Mars. One of the goals is to determine whether there are any forms of life on Mars. In 1976 NASA sent two Viking landers to Mars, but they found no signs of life. In fact, the conditions on the surface of Mars were so harsh that scientists gave up all hope of ever finding life on the red planet, and little time was devoted to finding foolproof methods of detecting life in space. This changed when evidence that water once flowed on Mars was obtained and when extremophiles were found on Earth. Extremophiles are extremely primitive organisms that flourish in extreme environments where there is no oxygen, for example, in near-boiling sulfurous waters found in deep-sea hydrothermal vents. These are the type of organisms that might have lived on Mars. Suddenly, it seems possible that there might have been life on Mars, and a large amount of effort is being spent on trying to detect lifeforms there, especially if they occur at very low concentrations. One of the main problems researchers are facing is that they have to make sure their ultrasensitive life-detection systems are able to distinguish between forms of life from Earth and those actually found in space. Imagine how embarrassing it would be if astrobiologists announced that they had found a new form of life, only to discover that some intrepid cold virus from Earth had hitched a ride on the Mars explorer.

Lots of research on sterilizing spacecrafts before takeoff was conducted in the sixties and early seventies, but that all stopped when no signs of life were found on Mars in the 1976 expeditions. According to Bob Koukol of NASA's planetary protection group, "Current technologies are basically the same as they were in the sixties and seventies."[14] One new technique that is being used by both the American and European space agencies is the use of firefly luciferase ATP assays to establish the amount of pre-takeoff contamination. It might even be used to determine whether there is life on Mars.

On Christmas Day 2003, the European Mars lander Beagle 2 went missing. The European Space Agency still hopes that it will hear the Beagle 2's calling sign—a nine-note tune composed by the British pop band named Blur—and that they can resume contact with the lander. The NASA missions were more successful, and we continue to get data from both Spirit and Opportunity, which landed on January 4 and 25, 2004, respectively. Both exploration rovers are trying to establish whether life was once possible on Mars. They have found signs that there was once water on Mars, but as of yet there has been no indication of life. If they find that life was possible, then they hope to employ luciferase tests using ATP to see whether there are any living organisms on the red planet.

These tests derived from the early work of McElroy. One of the questions that puzzled McElroy and Hastings is how fireflies control their sequence-specific light production. They had established that mixing all the ingredients required for light production in the lab resulted in a rapid flash of light that decayed in about twenty to thirty seconds, much longer than the two hundred millisecond flashes observed in the firefly. How did the firefly produce such short flashes? Hastings and McElroy found a clue to the mechanism when they mixed all the required ingredients in the absence of oxygen and no light was emitted. But, when they added oxygen to the mixture, they observed a burst of light that resembled that which is produced by the firefly. Based on these observations, they suggested that the light cells of the firefly were free of oxygen and that the length and sequence of their light flashes was regulated by a very precise oxygen release from the trachea that supply the light cells with oxygen. It was a great hypothesis, but McElroy and Hastings couldn't prove the mechanism or work out the details. In 1989 Hastings wrote, "The problem is still here, and it is a challenging one in cell biochemistry. Perhaps some of you will consider working on this problem again. Now, over 35 years later, with the many advances in instrumentation and knowledge of cell biology, some major new discoveries should be possible."[15]

Neurobiologists have observed that each flash produced by the firefly is accompanied by a burst of neural activity and the release of a primary neurotransmitter. This would imply that when the firefly feels the need to communicate its desire to mate, it sends out nerve signals to the light cells, which allow oxygen to join ATP, luciferin, and luciferase, and a flash of light is given off. Sounds fairly simple, doesn't it? Unfortunately, there is one big problem. The neurons that send the flashing signals from the brain to the light cells do not make it all the way to the light cells. They end just short of the photocytes. Somehow the message has to make that last jump from the end of the neuron to the light cell. In 2001 Barry Trimmer at Tufts University discovered that nitric oxide is produced in the cells that are located between the nerve endings and the light-producing photocytes and that it is most likely responsible for controlling the flashes in fireflies.[16]

The cells surrounding the photocytes are filled with elongated particles. Initially scientists thought that they were luminescent bacteria. However, it turned out that the elongated particles were mitochondria, the cell's power plants.[17] When the mitochondria are running at full speed, they use up a lot of oxygen, which has reached them through the tracheal air tubes described earlier. All mitochondrial power production comes to a screaming halt though when the nitric oxide is released. Now the oxygen can make its way to the light-producing parts of the photocytes, where it reacts with the luciferin-producing light. The light disappears as soon as the nitric oxide dissipates and millions of mitochondria start using up the oxygen again. This can all occur in a fraction of a second.

Nitric oxide is a very interesting molecule. It caused a scientific sensation when, in the early 1990s, it was discovered that this small molecule, previously known only as a toxic gas that is produced in car engines and after lightning strikes, had a crucial role inside living organisms. In humans it has many important functions, such as controlling blood pressure, penile erections, and the formation of memories. It was selected molecule of the year by *Science* magazine in 1992, and Robert F. Furchgott, Louis J. Ignarro, and Ferid Murad were jointly awarded the Nobel Prize in medicine in 1998 for their discoveries concerning "nitric oxide as a signaling molecule in the cardiovascular system." This superstar of the small molecule world is also responsible for flash control in fireflies.

We have seen how the luciferins of fireflies and other species give off light and how they can be used. Now I want to tell you about jellyfish bioluminescence and its story.

CHAPTER 4

SHIMOMURA'S "SQUEEZATE"

While Bill McElroy was studying the firefly, a very similar and yet quite different story was beginning to take shape. This is the story of the jellyfish protein that is responsible for its bioluminescence—the discovery of the protein called green fluorescent protein (GFP). This amazing protein has applications in many areas of medicine and biology and is even used in art. The story starts with Osamu Shimomura, the grandfather of the green fluorescent protein revolution. Shimomura was the first person to isolate GFP and to find out which part of GFP was responsible for its fluorescence. His meticulous research laid the solid foundations on which the GFP revolution was built.

Shimomura grew up in Japan during the Second World War.[1] His father was a professional soldier. As a child, Shimomura knew that he did not want to be a priest, medical doctor, or a teacher, but he did not know what he wanted to be. He wasn't particularly interested in science. Shimomura says he had never made a conscious choice to become a scientist; it was a series of coincidences, good fortune, and a bit of fate that led him to a life of investigating jellyfish bioluminescence.

There were not many opportunities for a college education in Japan immediately after the Second World War. Nevertheless, Shimomura was fortunate, as his high school teacher managed to organize a research assistantship for him at Nagoya University. It was to be both a job and an informal education. He worked with Professor Yoshimasa Hirata, who wanted him to isolate the material responsible for the bioluminescence of *Cypridina*, a tiny mollusk. When crushed and wet with water, *Cypridina* gives off a bright glow. Professor E. Newton Harvey and his research group at Princeton had been trying to isolate the bioluminescent material from the mollusk since 1916 without success. He was the doyen of bioluminescence research in the world at the time, and if he and his many graduate students could not do it, then it seemed probably impossible. Had Shimomura been a graduate student, he would never have been assigned this Herculean task and might never have gone on to study *Aequorean* bioluminescence. Fortunately, he was a research assistant. Since they are paid helpers, they can be assigned to work on "blue-sky" projects that have little prospect of success.

At first Hirata didn't tell his research assistant about the Princeton work or describe all the difficulties associated with the isolation. By the time he did break the news, Shimomura was so absorbed in the project he did not want to stop the difficult task and plowed ahead. It took more than ten years, but Shimomura finally managed in 1956 to isolate and crystallize the material that made *Cypridina* glow. Hirata and Shimomura published their results in 1959. Professor Frank Johnson, a faculty member at Princeton and a former student of Newton Harvey, read Shimomura's paper about *Cypridina* luciferin, the material that made it glow. He was extremely impressed by Shimomura's work and invited him to come to Princeton to work as his research associate. Shimomura and his wife, who had a pharmaceutical degree, agreed to come to America. Just before they left Japan for Princeton, Hirata arranged for Shimomura to receive a PhD based on the work he had done. It was, and still is, exceptional for someone to be awarded a PhD when he isn't enrolled in a doctoral program and hasn't even received an undergraduate degree. This is especially rare at a prestigious university such as the University of Nagoya. It was an amazingly generous gesture on behalf of Hirata—not that Shimomura didn't deserve it—one that had a tremendous influence on Shimomura's life and research, for it is difficult, perhaps nearly impossible, to get research funding in the United States without a PhD. It was also a reflection of the trust Hirata had in Shimomura's knowledge and ability.

In fall of 1960, shortly after Shimomura's arrival from Japan, Johnson asked him whether he was interested in studying the bioluminescence of *Aequorea*. After hearing Johnson's descriptions of the brilliant luminescence of the jellyfish and its abundance around Friday Harbor in the Puget Sound in Washington state, Shimomura signed on. Today, forty-three years later, he is still studying *Aequorea*'s luminescence.

In the early summer of 1961, Johnson, his assistant Yo Saiga, Osamu Shimomura, and Shimomura's wife packed Johnson's station wagon up to the roof and drove from Princeton to Friday Harbor. It must have been quite a road trip. As the only driver, Johnson was behind the wheel of the car for twelve hours a day for seven days—a car jam packed with three other passengers, a two-cubic-foot photometer, lab equipment, chemicals, and baggage. Together they made their way through Chicago and the Glacier National Park to Friday Harbor.

Upon their arrival, they were assigned a workspace in Lab 1, a small building with two rooms. There were three other scientists in the space, and one was Dr. Dixie Lee Ray, the future governor of Washington state. Even though dogs were not allowed in the laboratory, her dog was never far from her side. She declared that it was her assistant, not a dog. It seemed to work, as her dog was never thrown out.

Friday Harbor was the ideal spot to study *Aequorea aequorea* because there was a constant stream of jellies alongside the dock, riding the tidal currents. Despite the presence of so many jellyfish, it wasn't easy to catch them because they are transparent in the water. This is an excellent defense mechanism. *Aequorea* only fluoresce when they are agitated or are placed in a potassium chloride solution. A typical medusa from Friday Harbor was rather large, with a diameter of three to four inches and a weight of fifty grams. It was fairly strenuous to catch and lift hundreds of the sizable jellyfish out of the water. After they were caught, scissors were used to cut off a thin strip at the edge of the jellyfish's umbrella. This ring contains a hundred or so evenly distributed light organs that produce fluorescent green light.

Fortunately, cutting *Aequorea* did not hurt them, as they have no central nervous system. When the rings of twenty to thirty jellyfish were squeezed through a rayon gauze, a faintly luminescent liquid called *squeezate* was obtained. Johnson and Shimomura had come to Friday Harbor to collect this squeezate and to extract from it the substance responsible for its luminescence. After a few days of work, they had tried every method of isolating the luminescent material they could think of, but with absolutely no success. The chemistry of the firefly luminescence was well known at the time, and their methods were all based on the assumption that the luminescence in the jellyfish would be similar to that observed in the firefly, or *Cypridina*. They had set their sights on a luciferin and a luciferase system. Shimomura then suggested that this might be the reason for their inability to separate the luminescent material from the squeezate, but Johnson was not convinced, so the two started working at separate tables. Neither had much success. Shimomura spent several days analyzing the problem. His favorite place to do his thinking was on a rowboat, drifting in Friday Harbor. He felt safe in his boat, as rowboats always have the right of way over motorized vessels. Even the large ferries gave him a wide berth. The only drawback to his rowboat meditation was that if Shimomura's thoughts were too strenuous and he fell asleep, he had to row for a long time to get back to the Friday Harbor laboratory. Amazingly enough, thinking on the rowboat did the trick, and inspiration came while Shimomura was bobbing up and down in the harbor.

One of the things he was looking for was an on/off switch for the squeezate luminescence. While on the rowboat, he decided that a protein was most likely responsible for emitting the luminescence observed coming from the jellyfish light organs. If he were right, then he could stop the light emission by preventing the protein from giving off the light. Proteins are pH sensitive (pH is just a measure of acidity), and so by changing the acidity of the squeezate, Shimomura hypothesized that he could switch off the luminescence. If he were really lucky, he could turn it back on by returning to the original pH. Chemists call this *reversible inactivation*.

Shimomura immediately rowed back to the lab and prepared some fresh squeezate. Using acetic acid, he adjusted the pH of three samples of the squeezate to pHs of 6, 5, and 4. All the hard thinking in the rowboat paid off. The two solutions with a pH of 6 and 5 were luminescent, but the one with a pH of 4 was not. Shimomura filtered the squeezate solution that had a pH of 4. The resultant liquid, free of cells and other jellyfish debris, was nearly dark, but its luminescence returned after adding sodium bicarbonate to neutralize the acid. Shimomura was understandably excited. He had demonstrated that a protein was probably involved in jellyfish luminescence, and, more importantly, he had shown that the protein was in the neutralized solution—it could be isolated from the jellyfish. However, Shimomura's day was not over yet—the best was yet to come. When he added a little seawater to the solution, he saw it burst into luminescence. Blue luminescence! Something in the seawater was activating luminescence in the squeezate solution. But why was the solution giving off blue light, when the light organs of *Aequorea aequorea* were fluorescent green? It didn't take Osuma Shimomura long to realize that it was the calcium ions in the seawater that were responsible for the increase in luminescence, but it would take a lot longer before he could figure out why his extract's luminescence was blue while the *Aequorea*'s was green.

Shimomura's was one of the most important experiments in the green fluorescent protein story, and the amazing thing is that it was basically a kitchen experiment. Shimomura used squeezate, vinegar (acetic acid), baking soda (sodium bicarbonate), and seawater. On the basis of this experiment, Johnson and Shimomura were soon able to isolate the protein responsible for luminescence in the jellyfish. Unfortunately, the amounts of protein they isolated were miniscule. It was July 1961, and there was no more time for soul-searching rowboat rides. Many more jellyfish rings were required before the Princeton researchers would have enough of the luminescent material to characterize it. In order to catch and process as many jellyfish as possible, they started a jellyfishing routine. Typically, they would collect jellyfish from six to eight a.m., have a quick breakfast, and then cut the umbrella rings until noon. The afternoons were spent making squeezate and extracting the luminescent material. This was no summer vacation, and from 6:30 to 8:30 p.m. more *Aequorea* were collected. They were placed in an aquarium, where they were kept overnight and processed after breakfast with the next morning's catch. It took at least one minute to cut each jellyfish ring, and four people working for three hours could not produce more than five hundred rings. That wasn't enough, and since these jellyfish don't sting, high school girls were trained to cut rings; they were paid two cents for each ring they cut.

Johnson and Shimomura also bought jellyfish from children of scientists living on the campus for a penny per jellyfish. Unfortunately, the jellyfish disappeared from Friday Harbor before they had enough protein. This was not surprising, since *Aequorea aequorea* live for six months or less, and the entire population dies by mid-autumn every year. New generations of medusae return each spring. If Johnson and Shimomura wanted more of the luminescent material, they would have to come back the following spring and harvest a new generation of *Aequorea*. The jellyfish extract of "only" ten thousand specimens was packed in dry ice and driven back to Princeton.* After six months of painstakingly careful work, five milligrams (0.005 g) of pure protein was obtained. The protein gave off blue light when a trace of calcium ions was added. In honor of *Aequorea aequorea*, they named the protein *aequorin*. In 1962 they published a paper describing the isolation, purification, and properties of aequorin; it contained the following footnote: "A protein giving solutions that look slightly greenish in sunlight though only yellowish under tungsten lights, and exhibiting a very bright greenish fluorescence in the ultraviolet of a Mineralite, has also been isolated from the squeezates."[2]

Although the existence of a green fluorescent substance in *Aequorea* was known since 1955, this was the first time that the substance was identified as a protein. Proteins are an essential part of every living cell. They transport oxygen, nutrients, and minerals through the bloodstream, and they are also major components in skin and muscle. Many of our hormones that act as chemical messengers are proteins. Shimomura had shown that the material was a protein. Seven years later, the protein would be given the very appropriate name *green fluorescent protein*.

The GFP revolution started small. GFP was only alluded to as a footnote in the aequorin paper, and only a tiny amount of what would later be called green fluorescent protein had been isolated as an offshoot of the aequorin project.

From 1962 to 1970, Shimomura concentrated his research on aequorin and its structure. GFP was only interesting to Shimomura because of its biochemical links to aequorin. It was responsible for the live jellyfish giving off green luminescence, while aequorin isolated from the same jellyfish emitted blue light.

---

*Shimomura's word.

In order to do his research, Shimomura estimates that he collected over a million *Aequorea* specimens, cut off the rings, produced squeezate, and isolated GFP and aequorin. The jellyfish were caught from four different docks, and a car was used to transport the buckets filled with *Aequorea* back to the lab. When the currents were uncooperative and carried the jellyfish away from the docks, rowboats were used. This was a tricky activity that left a number of jellyfishers in the cold Pacific waters. Jellyfish don't always just float with the current; they can swim when needed.

One of the earliest reports of jellyfish propulsion is by Alexander von Humboldt. In his *Personal Narrative of Travels to the Equinoctial Regions of America, during the Years 1799–1804*, he wrote,

> We looked in vain for sea-weeds and mollusca, when on the 11th of June we were struck with a curious sight which afterwards was frequently renewed in the southern ocean. We entered on a zone where the whole sea was covered with a prodigious quantity of medusas. The vessel was almost becalmed but they were borne towards the south-east, with a rapidity four times greater than the current. Their passage last near three quarters of an hour. We then perceived but a few scattered individuals, following the crowd at a distance as if tired with their journey? Do these animals come from the bottom of the sea, which in these latitudes some thousand fathoms deep? Or do they make distant voyages in shoals?[3]

Nearly two hundred years after von Humboldt's observations, researchers have shown that not all jellyfish swim in the same way.[4] By examining six jellyfish species, Sean Colin at the University of Connecticut, Avery Point, and John Costello from Providence College were able to determine that flat-shaped jellyfish use their umbrella edges to row through the water, while the squid-shaped jellyfish use jet propulsion. The jet-propelled jellyfish are ambush predators. They spend most of their time floating around, waiting for prey to get entangled in their extended tentacles. They retract their tentacles when they have to move and therefore cannot feed while swimming. Jet propulsion is rapid but requires a lot of energy. Perhaps that is why the ambush predators are not very active and swim only when threatened or when repositioning themselves in the water column.

The rowers are cruising predators and swim far more than the ambushing jellyfishes. By examining the movement of naturally occurring phytoplankton in the wake and vortices of the "rowing" jellyfish, the investigators were able to show that the rowing action creates a feeding current that draws prey past the edge of the medusa into the trailing tentacles. Cruising predators have to swim a lot in order to catch as much prey as possible; that is why they use the more efficient yet slower rowing motion. *Aequorea victoria* is one of the cruising predators that uses a rowing mode of propulsion. (See figure 3 in the photo insert). Its preferred diet is other jellies, and it can eat jellyfish up to half its size.

The whole Shimomura family—wife, husband, daughter, and son—were often accompanied by a couple of assistants on their jellyfish-collecting expeditions. On good days when a dense stream of medusae was passing alongside the dock, they could collect five to ten jellyfish per minute. These were placed in buckets filled with seawater. After cramming about one hundred jellyfish in a bucket, it would be considered full and taken to the trunk of the car, which had room for twelve buckets. Once the car was filled, the jellyfish were rushed to the lab where they were released into aquaria. On average about three thousand to four thousand jellies were collected per day. The town dock, a public pier, was an excellent site for collecting jellyfish, but there were some problems, the biggest of which was the large amount of time spent answering questions: "What are you collecting?" "Why are you collecting jellyfish?" Most passersby were satisfied with a simple reply, such as "These are for scientific research," but some pressed for a longer, more detailed description. Shimomura recalls an incident when an old lady poked her head out from the window of a small boat, looked at the jellyfish on his net, and asked him, "How do you cook them?"

Shimomura answered, "I don't cook them."

She looked contemptuously at him and asked, "Do you eat them raw?" Before he had a chance to answer, she withdrew back into her boat.

By 1969 the Shimomura group had developed two ring-cutting machines that were able to cut off the light organ–containing rings ten times as fast as they were able to do by hand. Now a practiced team of two could cut and extract squeezate from 3,360 jellyfish in two hours and forty minutes. By the early seventies, Johnson and Shimomura had enough squeezate to purify about a hundred milligrams of GFP. It had a beautiful bright lime-green color, and when it was irradiated with ultraviolet light, it emitted a fluorescent green glow. Frank Johnson wanted to take some of the GFP Shimomura had isolated from *Aequorea* to a bioluminescence conference in Monterey, California, so that he could show his colleagues the beautiful material. Shimomura gave him all the GFP they had extracted up to that point. Johnson returned from the conference terribly embarrassed; he had handled the sample as if it had been his own baby, but somehow he had lost it. Shimomura and his wife had to start all over again. Frank Johnson had been so good and helpful to them that Shimomura immediately forgave him for losing their GFP supply.

In the early seventies, Johnson and Shimomura reported that the calcium-triggered luminescence of aequorin was blue in the absence of GFP but that at high GFP concentrations, the luminescence of aequorin is absorbed by GFP and emits green fluorescence. Furthermore, "Even in a crude preparation, GFP can be easily recognized by its bright green fluorescence. . . . The fluorescence is so bright that it can easily be recognized in ordinary room light."[5]

A big question that still remained unanswered was why GFP was fluorescent. Proteins are found in all living organisms, and they are everywhere in the human body: in the hair, blood, muscle, and the thousands of enzymes that regulate the chemistry of the body. None of these proteins is fluorescent. It was more than ten years after Shimomura had shown that GFP was a protein, but he still couldn't figure out why it was fluorescent.

All proteins are long strings of amino acids, and there are only twenty amino acids. It is amazing to imagine that the only difference between all the proteins, with their large variety of functions, is the number of amino acids composing the chain and the order of the amino acids. A protein is somewhat like a word. Words are composed of letters, and the only differences between all the words in the English language are the number and the order of the letters. What is it about the order and the number of amino acids in GFP that make it glow? The first thing Shimomura did to answer this question was to take an enzyme that cut up proteins whenever it came across certain combinations of amino acids, and he used it to cut GFP into smaller strings of amino acids. He separated the fragments and examined their response to light to find out which one was responsible for

GFP's fluorescence. None of the fragments glowed. However, one of the fragments absorbed light in exactly the same way as GFP did. Shimomura correctly assumed that it must be this fragment that was responsible for GFP's fluorescent behavior. In a brilliant piece of detective work, he determined the structure of the protein piece responsible for GFP's fluorescent behavior. It took less than a month and was largely based on the fact that Shimomura recognized that the luminescent behavior of the fragment was very similar to the *Cypridina* luciferin structure he had established in 1956. The structure of the GFP chromophore, that is, the fragment responsible for GFP's fluorescence, was published in 1979.

In the same year, William Ward from Rutgers University came to visit Shimomura and asked him whether he minded if the Rutgers group worked on GFP. Shimomura was more interested in aequorin than GFP and agreed to stop working on GFP and concentrate on aequorin. That way, there would be no overlap in their research efforts. This is a fairly common practice in science. Rather than duplicate their research, groups often divide research tasks among themselves.

Most of this book is devoted to GFP, but this is a good place to take a small detour and write a bit about aequorin. There are two reasons for doing this. First, aequorin was Shimomura's baby, and it's interesting to see what he found so fascinating about this protein. Second, aequorin has been used as a luminescent indicator for calcium for more than thirty years. It is protein that consists of 189 amino acids. It needs to bind to something called *coelenterazine*, oxygen, and calcium before it glows. When three calcium ions bind to aequorin, it changes its shape, which results in oxygen being able to react with coelenterazine to form a compound called *coelenteramide*, which is responsible for the emission of blue light. However, without calcium binding the aequorin, no blue light is given off; therefore, the emission of blue light by aequorin is a very good indicator of the presence of calcium. This means that aequorin could be used as a probe for calcium in much the same way as luciferase has been used to detect ATP.

Humphry Davy, mentioned earlier in chapter 2, was the first to discover the existence of the element calcium in 1808. He named the element after the Latin word for lime, which is *calx*. About 3 percent, by weight, of the Earth's crust is calcium; most of it is sedimentary rock.

Typically rainwater contains very little calcium, freshwater has about ten times as much calcium as rainwater, and seawater has approximately thirty times as much calcium. Water that contains high concentrations of calcium is called *hard water*. Beer contains more than ten times as much calcium as freshwater does, and it is claimed that the taste of the beer strongly depends on the calcium concentration.[6]

It wasn't long after Davy reported his discovery that the important role of calcium in the formation of mammalian bones and other mineralized tissue was found. Today, many people take calcium pills to supplement the amount of calcium they ingest in their daily intake of milk, cheese, yogurt, and other calcium-containing foods. This is mainly done to prevent osteoporosis in the later years of life. Human bone is constantly being made and broken down. Osteoporosis occurs when the rate of breakdown exceeds that of formation, which significantly reduces the strength of the bones. About 50 percent of American women and 25 percent of American men over forty-five are affected by osteoporosis.[7] The United States' recommended daily allowance for calcium for adults is eight hundred milligrams. During pregnancy it is extremely important that calcium is transported across the placenta, especially during the developmental stages, when bone growth is most rapid.

In 1883 Sidney Ringer accidentally found out that calcium, which was present in high concentration in the London tap water of the time, was required for muscle contraction and tissue survival in frog heart muscle. Since then, calcium has been shown to have a central role in the transmission of messages within cells, messages that regulate important processes such as secretion, cell division, muscle contraction, glucose formation and breakdown, and growth. Calcium also has significant functions outside of cells. In blood plasma of mammals, it plays a central role in the formation of blood clots, and, in plants, it forms links between cells, providing rigidity.[8]

Aequorin is an excellent protein for measuring the concentration of free calcium ions in living cells. Prior to 1991, jellyfish aequorin and coelenterazine were microinjected into cells in order to measure their calcium concentrations and the concentration changes that occur inside them. If calcium were present in the cells, the response was rapid and could be accurately measured by means of image intensifiers and photon counting. The intensity of the flash of light produced in the presence of calcium is proportional to the amount of calcium present.[9] In the early nineties, aequorin was cloned. This means that the genetic material required to make aequorin in jellyfish was removed from a jellyfish cell and multiplied so that it was now possible to take the recipe (gene) for making aequorin and insert it into other cells, which would then make aequorin. By placing the foreign aequorin gene in the correct location in the genome of the host cell, it was now possible to produce cells that would make jellyfish aequorin in the organelles of interest.

For example, a modified plant cell could be produced in the lab that made aequorin in its cytosol, which is the fluid inside a cell. Incubation of this plant cell with coelenterazine would result in a flash of light if calcium were present. You might ask why would someone want to measure calcium in plants. That would be a good question. The answer is an interesting one that would probably surprise you—it certainly astonished me. Plant intelligence is the answer; cellular calcium is critical to a plant's intelligence. Yes, plants can "think"; they continuously screen at least fifteen environmental variables and by using calcium signals adapt to these external stimuli.[10] The stilt palm, for instance, just ups and moves away when competitive neighbors approach it; it can do this by differential growth of the prop roots that support the stem. The parasitic dodder plant is a lot more aggressive than the stilt palm. Within an hour of its first contact with a prospective host, it assesses its potential. If the dodder "decides" that the host doesn't have enough to offer, it searches for a new victim. Once a suitable host has been found, the dodder encoils its host; the number of coils and suckers it uses depends on its assessment of the host's nutritional potential. Like all plants, the dodder does not have a brain, nor does it have a central nervous system. Cellular calcium mediates most plant signals, but we have no idea how this occurs.[11]

Anthony Trewavas at the Institute of Cell and Molecular Biology at the University of Edinburgh has often used aequorin to examine how plants react to external stimuli, such as wind. Plant growth and development is dramatically influenced by wind. Crop yields, plant size, and leaf area are reduced even after short exposures to wind. In order to establish how plants perceive wind signals, Trewavas and his coworkers genetically transformed some tobacco plants so that they made aequorin in their cytosol. Treatment of these plants with coelenterazine resulted in luminous plants whose light emission was directly proportional to the amount of calcium present in the cytosol.[12] Using these plants, the authors showed that wind stimulation caused an immediate increase in cytosolic calcium. The increases occurred only when the plant was in motion, indicating the possibility that calcium signaling is involved in the plant growth response to wind. The plants were also exposed to a number of other external stimuli. Cytosolic calcium concentrations increased in response to touch and cold shock, but not to heat shock or chemical exposure.[13] Trewavas has used these findings to design new tobacco plants that will function as early warning signs to farmers that their plants are under stress. The project has been partially successful. Wounded, infected, or otherwise stressed plants responded by releasing calcium ions, which bound to aequorin and caused the coelenterazine to give off a very faint blue glow that could be detected by ultrasensitive camera equipment. Up to now, they have been unable to create plants that will glow bright enough so that their stress is visible to the naked eye. Ultimately, Trewavas and his coworkers hope that their stress alarms will make the widespread use of blanket insecticide-spraying practices less prevalent. With luminescent aequorin-based, calcium-sensing plants acting as early warning systems, farmers might be able to limit their spraying to affected areas and spray in time to rescue their harvests.[14]

Let's look at two more applications using the aequorin gene to measure the amount of intracellular calcium.

Diatoms are extremely common in both freshwater and marine ecosystems. Their cell walls form exquisitely beautiful silica shells that were once one of the favorite objects viewed by early microscope users. The very same shells are also evident in fossils going back to the Cretaceous period. The vast majority of diatoms float freely, though some are found adhering to snails, crabs, turtles, and even whales. They are a major component of marine plankton and a major food resource for marine and freshwater microorganisms and animal larvae. They contain chlorophyll and are capable of photosynthesis, which means that they use water, sunlight, and carbon dioxide to make oxygen and sugars. It is estimated that diatoms are responsible for removing between 20 and 25 percent of all carbon dioxide from the atmosphere. Since carbon dioxide is a greenhouse gas, diatoms are crucial to our existence on Earth. Despite their obvious importance, we know very little about diatoms. No one knows what diatoms can sense in their environment, how they respond to external signals, and what factors control their life strategies. Historically it has always been assumed that plankton are passive and incapable of responding to external stimuli. But that view is changing, in part due to studies using aequorin to detect changes in the calcium concentration of diatoms. Since 1999 it has been possible to generate transgenic diatoms—these are diatoms with foreign genes in them. One of the first transgenic diatoms created was the marine diatom *Phaeodactylum tricornutum* containing a gene for aequorin. They were used to analyze diatom responses to osmotic stress, fluid motion, and iron—which is a key nutrient that controls the amount of diatoms present in the ocean. Changes in the calcium concentration were observed in response to all three stimuli. Therefore, it was concluded that diatoms can detect and respond to physical and chemical changes in their environment using sophisticated perception systems that are based on changes in their calcium concentration.[15]

The last use of aequorin calcium indicators to be described here is the examination of circadian phases. I have always been fascinated by the internal clock we all have. In my twelve years of teaching at college, I have seen numerous students whose circadian rhythms have been off a little, and when I was doing my postdoctoral studies, there was a graduate student in the laboratory who had immense difficulties getting up before noon. He would be the happiest when he came to lab in the mid-afternoon and worked through the night, going to bed at four or five in the morning. Recent studies have shown that as children reach adolescence, their circadian rhythms shift and they work most efficiently when they go to bed late and sleep longer. Some high schools in Connecticut have shifted their school day to accommodate these findings. Humans aren't the only organisms that have an internal clock. In fact, nearly all living organisms, from fruit flies and butterflies to plants, have circadian rhythms, which continue running in total darkness and the absence of external stimuli. Circadian phases are responsible for such diverse phenomena as jetlag and wayfinding in monarch butterfly migrations. These phases are most commonly entrained by daily changes in light but can also be linked to changes in humidity and temperature. It has been shown that bacteria, whose circadian clocks are synchronized to the light/dark cycle of their environment, exhibit an enhanced fitness. How do the cells in an organism get the "time" from their internal clock? Is there a central clock that sends out time signals to all the surrounding cells, or do individual cells have their very own internal clock? Cytosolic calcium ions are a messenger in plant cells—could they be responsible for sending circadian information within and between cells? All these questions have been answered, for plants at least, using transgenic plants containing aequorin. Scientists have demonstrated that red light, which can be used to train the circadian clock, leads to temporary increases in cytosolic calcium concentrations, which undergo clearly visible circadian variations, indicating that calcium is used to signal the time in the aequorin-labeled cells. They also found that different cell types light up in different phases; they have different clocks regulating their internal functions.[16]

Transgenic tobacco plants containing both the jellyfish aequorin and the firefly luciferase gene have been generated to examine the possibility of separate circadian pacemakers controlling molecular events in plants. Tobacco plants and seeds produce a series of light-harvesting proteins in a well-characterized circadian rhythm. Researchers at Vanderbilt University in Tennessee have fused firefly luciferase genes to those for the light-harvesting proteins to produce transgenic seedlings that luminesce when light-harvesting proteins are produced. The researchers also added the aequorin gene to the seedlings in order to see whether the rhythm of the calcium-ion releases and the light-harvesting proteins are linked. Under natural conditions, the circadian phases are determined by light/dark and temperature signals so that they are both synchronized. Under red constant light, the two phases disassociate and oscillate with significantly different periods. Furthermore, the light-harvesting proteins continue their robust circadian rhythms in the absence of calcium-ion oscillations. In this way, glowing genes from the jellyfish and firefly in tobacco seedlings have been combined to show that there are at least two different circadian pacemakers that control molecular events in plants.[17]

Now let's leave aequorin and go back to Osamu Shimomura and green fluorescent protein. After arriving in America in 1960, the Shimomura family stayed at Princeton for the next twenty years. During all this time,

Osamu Shimomura worked as a research associate. His salary was paid by research grants mainly from the National Science Foundation. It was a tenuous situation, and in the early eighties, the biology department at Princeton reportedly decided to stop providing Shimomura with lab space. He thereupon moved to the Marine Biology Laboratories at Woods Hole, Massachusetts, where he stayed until his retirement nearly twenty years later. When I went to interview Osamu Shimomura at his home in Woods Hole in May 2003, he met me in his basement laboratory. It was a neat and obviously well-used lab with modern high-tech equipment. Although he was retired, Shimomura was still doing research. He was a charming and forthcoming host. The first thing he did was show me a flask of *Cypridina* that had been collected in Japan in 1944. The shells had survived nearly sixty years. Shimomura took a handful of the mollusks and crushed them. He switched off the lights and wet his hands with tap water. Gradually his hands began to glow, and soon they emitted light that was brighter than the glow-in-the-dark necklaces sold at fireworks displays. The Japanese had planned to use this luminescence to read maps and follow each other in the dark jungles during the Second World War. The plan failed when torpedoes sank the ships containing dried *Cypridina*, and the samples were ruined by humidity. After the war, the US Navy donated the remaining *Cypridina* from Japan to the Princeton lab. They were the source of the luminescence covering Shimomura's hands. I was also shown the luminescence of a sample of aequorin, and for the first time in my life, I saw some "real" GFP obtained from about fifty thousand jellyfish. All the GFP I had seen before was cloned, produced outside of living jellyfish.

The story of Osamu Shimomura's life is a remarkable one. He claims his research breakthroughs in bioluminescence were the result of a string of coincidences. That might be true, but I am sure that if fate had led him on a different path, he would have had an equally productive career. Shimomura has the right combination of perseverance, intelligence, and insight to lay the seeds of the GFP revolution.

A bit more than fifty years ago, James Watson and Francis Crick reported the double helical structure of DNA. It was a very different scientific breakthrough. Their ideas were based on published information, model building, and a fair amount of brainstorming conducted—this time in a pub, not a rowboat. If they hadn't discovered the double helical structure, Linus Pauling or Rosalind Franklin would have likely come up with the structure. The information was available, many research groups had generated it, and it was a race to interpret all the data. I doubt that GFP would be used in biotechnology today if Shimomura had not undertaken the tedious, painstaking, and meticulous process of isolating GFP from *Aequorea victoria*.

Interestingly, Shimomura's children, Tsutomu and Sachi, who grew up catching jellyfish and helping Osamu Shimomura with his research, seem to have the same intellectual curiosity and perseverance that characterizes their father. On Christmas Day in 1994, someone hacked into Tsutomu Shimomura's computer and stole thousands of files from him. Osamu's son did not take this intrusion into his private life lying down—he followed the intruder's path into his computers and monitored his Internet activities. He found that the hacker was also stealing computer files from Motorola, Apple Computer, and twenty thousand credit-card account numbers from a commercial computer network. Tsutomu Shimomura was a computer security consultant for the FBI and kept the agency informed of his findings. He suspected that Kevin Mitnick was to blame. Mitnick was a well-known hacker and on the FBI's list of most wanted criminals. Like his father, Tsutomu used his intellect and work ethic to trace Mitnick's cyber movements, at times spending thirty-six-hour shifts monitoring Mitnick's Internet activities. His efforts were rewarded, and in February 1995, the FBI and Tsutomu caught Mitnick in Raleigh, North Carolina. The story made the headlines of most newspapers and magazines and has been reported in three books.

Sachi Shimomura, Osamu's daughter, is no slouch either. She majored in English and mathematics at Stanford, completed her doctorate in medieval English at Cornell, and now teaches at Virginia Commonwealth University.

Osamu Shimomura is currently writing a textbook on bioluminescence. He predicts it will take him three years to complete and hopes that it will provide the impetus for two or three young students to concentrate on bioluminescence and answer some of the multitude of questions that remain unanswered. Shimomura is still extremely interested in the basic science of bioluminescence, but not in its applications. He is very surprised by how popular and useful GFP techniques have become, but he is concerned about gene manipulation and thinks it is dangerous to change nature. He is disconcerted by the behavior of modern scientists who are more interested in finding commercially viable applications than in discovering their underlying chemistry. One could say that Osamu Shimomura is the reluctant grandfather of the GFP revolution.

CHAPTER 5

WHERE IS THE GFP RECIPE? LET'S PHOTOCOPY IT

Douglas Prasher was the first person to foresee the potential for a GFP revolution, but not even in his wildest dreams could he have predicted the extent and impact it would have on the world of biotechnology and medicine. There was no way he could have imagined that the GFP gene from the jellyfish would one day be inserted into tadpoles so that they would fluoresce in the presence of heavy metals or that a pig would be cloned with a yellow snout formed by placing a gene of a modified form of GFP into its DNA.

In 1987 Prasher first got the idea that sparked the GFP revolution when he thought that GFP from a jellyfish could be used to report when a protein was being made in a cell. Proteins are extremely small and cannot be seen, even under an electron microscope. However, if one could somehow link GFP to a specific protein, for example, hemoglobin, one would be able to see the green fluorescence of the GFP that is attached to the hemoglobin. It would be a bit like attaching a lightbulb to the hemoglobin molecule.

In order to understand how the jellyfish GFP could be used to show when hemoglobin is made and where it goes, we need to know something about DNA, genes, and proteins. The molecular machinery in the cell makes proteins by using instructions encoded in the DNA of the cell. One can think of the DNA as a large cookbook that contains all the recipes required to make every kind of protein found in the body. Scientists call the recipes *genes*. Every cell has all the recipes required to make every protein that is found in the body. Since proteins are long strings of amino acids, each gene is a set of instructions specifying the order of all the amino acids in the protein. When a cell in your finger needs to make more muscle, the molecular machinery of that cell will find the recipes for the required proteins and make them. Unless something goes wrong, that cell will never make a corneal protein or any other protein that isn't needed in the finger. The complete set of instructions on how to make all the proteins in the body is called the *genome* (in essence, the recipe book). Humans are made up of about ten million million (10×10^{12}) cells, each cell has a nucleus, and in each nucleus lies a complete set of instructions. If one could stretch the DNA from one human cell in a straight line, it would be about two meters long. Taking the entire DNA from one person would produce a DNA string that could stretch from the Earth to the sun and back again more than sixty times.

With this knowledge we can now go back to the hemoglobin example. Hemoglobin is a vital protein in the body, since it carries the oxygen in our blood. Our bodies are continuously making new hemoglobin. Encoded in the DNA is some type of index that directs the molecular machinery to the start of the hemoglobin gene. When new hemoglobin is required, protein production is activated. The gene is read and the protein is manufactured. At the end of the gene is a message called a *stop codon*, which ends protein production. The manufacture of proteins using the instructions from the gene is called *protein expression*.

Doug Prasher envisioned that it would be possible to use biomolecular techniques to insert the GFP gene at the end of the hemoglobin gene, right before the stop codon. When the cell needed to make hemoglobin, it would go to the hemoglobin gene, use the information encoded in the gene to make it, but instead of stopping when the hemoglobin was made, this cell would carry on making GFP until it reached the stop codon at the end of the GFP gene. As a result, the cell would produce a hemoglobin molecule with a GFP attached to it.

There were three reasons Prasher thought that GFP could potentially become a significant tracer molecule. First, if enough protein with attached GFP were made, it should be easy to detect and to trace it as it moved through the cell because irradiating the cell with ultraviolet light would cause the GFP attached to the protein to fluoresce. Second, Shimomura had shown in 1974 that GFP was a fairly small protein. This was important because a small protein attached to the protein of interest was less likely to hinder its proper function. Its small size would also allow it to follow the fused protein, especially in organelles like neurons, whereas the diffusion of large proteins would be difficult. Third, it had been shown that once GFP was made in the jellyfish, it was fluorescent. Most other bioluminescent molecules require the addition of other substances before they glow. For example, aequorin will glow only if calcium ions and coelenterazine have been added, and firefly luciferase requires ATP, magnesium, and luciferin before it luminesces. This would make GFP a much more versatile tracer than either aequorin or firefly luciferase, which were being used as tracers.

Prasher's only concern was that it was quite likely that fluorescent GFP could be formed only in the jellyfish because there were other proteins that helped GFP form the region of the protein responsible for fluorescence, the *chromophore*, in the jellyfish that were not present in other organisms.

Besides attaching GFP to a protein and making it a fluorescent tag, Prasher also thought that GFP could potentially be a very useful reporter molecule. In order to activate protein production, DNA promoters are used; these are sequences of DNA next to genes that contain the information about where and when the gene should be read and make the protein (in other words, be *expressed*). If GFP is linked to a specific promoter, then it will be expressed in place of the protein, showing where and when the gene of interest is switched on. To go back to the cookbook analogy, if we wanted to know what happens to a lemon meringue pie after it has been made, we could add a few lines to the end of the recipe that request the addition of a permanent fluorescent dye. The dye would be our tracer. If the pie were placed in the midst of many others, we would still recognize it because of its fluorescence. When GFP is used as a reporter molecule, we aren't really interested in the fate of the molecule being produced—we want to know when it is made and where it is made, not what will happen to it.

Although he realized that GFP had the potential to be a very useful reporter and tracer molecule, Prasher was cautious. Before GFP could be used as a fluorescent tracer molecule, the location of its gene needed to be found. Furthermore, even if he could find the GFP gene in the jellyfish, Prasher knew that there was no guarantee that the GFP gene expressed in organisms other than the jellyfish would produce fluorescent GFP.

Before we go on to see what Douglas Prasher did to meet this challenge, let's first meet the man.

If you were able to go back magically in time to Douglas's youth, I think you would be surprised that this was the person who would give birth to a revolution in biotechnology. For Douglas was well into his university career before he took his first biology class.[1] He was born outside of Akron, Ohio. When he was in his teens, his parents moved to Ontario, Canada. There, due to the difference in the schooling systems between Ohio and Ontario, Douglas went through high school without ever taking a biology class. He enjoyed the physical sciences and decided to go to Ohio State University to major in chemical engineering, but he soon discovered that he wasn't enjoying his classes. Pretty much on a whim, he changed his major to biochemistry and started taking biology classes. Douglas enjoyed the subject so much that after graduating with his bachelor's degree, he decided to stay in the biochemistry department at Ohio State University to complete his PhD in 1979. Douglas then went to the University of Georgia to do a postdoc with Sidney Kushner, a bacterial genetist. It was during his postdoc with Kushner that Prasher first met Milt Cormier, a professor at the University of Georgia, with an interest in bioluminescence. Prasher found the work very interesting and changed groups after he was done with his first postdoc. During his time as a postdoctoral assistant and assistant biochemist in Milt Cormier's group, he found the location of the aequorin gene and managed to isolate it from the *Aequorean* DNA. Most of his time in Cormier's group was spent working on aequorin. But it was with Cormier that Prasher developed an interest in GFP, one of the proteins he was working on as a side project.

In 1987 he moved from Georgia to Massachusetts, where he became an assistant scientist for the biology department of the Woods Hole Oceanographic Institution, Massachusetts. One of his research projects at Woods Hole was to locate the GFP gene, clone it, and sequence it.

Proteins are long strings of amino acids, and their sequence determines the properties of the protein. It's very similar to letters and words. For example, if we take the four letters *a*, *e*, *m*, and *t*, we can arrange them in twenty-four ways. Most of them don't make much sense in the English language, but some, like *tame*, *mate*, *team*, *meta*, and *meat*, do. The sequence of the letters determines the meaning and the function of the word.[2] Take

the morse code

rearrange the letters and we can get

here come dots

The same can be done with

two plus eleven,

which can be rearranged to give

one plus twelve.

While words are made up of a combination of twenty-six letters, proteins are made up of combinations of twenty amino acids. *Team* is always spelled T-E-A-M; in the same way, every GFP will have the same number of amino acids, and they will always be in the same order. Changing just one amino acid can change the properties of the protein. This is, in some ways, more impressive with proteins than with words because proteins comprise a string of hundreds of amino acids that have to be in the right order. Hemoglobin is a protein that carries oxygen around in the blood. It has 574 amino acids. They can be combined in 6×10^{746} different ways, which is six with more than a page of zeros behind it. Only one of the combinations is hemoglobin; changing just two of the 574 amino acids can result in a defective protein, which would result in sickle cell anemia. Normally, red blood cells are doughnut shaped, but in people with sickle cell anemia, the doughnut has collapsed, resulting in defective red blood cells that are inefficient oxygen transporters. So, in order to understand why GFP fluoresces, one needs to know its sequence.

Frank Prendergast of the Mayo Foundation for Medical Education and Research and Bill Ward of Rutgers University, who had started working on GFP after talking to Shimomura in 1979, found the sequence of a stretch of amino acids in GFP. Prasher used this information to try to find the gene that had the information to make the protein with the sequence of amino acids that Pendergast and Ward had found in GFP. This is the biomolecular equivalent of going through all the cookbooks in the kitchen to find the recipe that contains the fragment "beat three eggs slowly, add two tablespoons of mango chutney, and bake for ninety."

Prasher succeeded in finding a gene that had the information to make a protein containing the amino acid sequence that he was looking for. Unfortunately, the resultant protein was composed of only 168 amino acids, which was smaller than Shimomura had determined GFP to be. It was, in fact, the right protein, but the gene that Prasher had managed to isolate was incomplete; the code for some of the first and last amino acids was missing.[3] All Prasher's work wasn't completely wasted since the gene fragment he isolated contained the instructions for the region of GFP that was responsible for some of its luminescent behavior. Later this fragment would be used to confirm the structure of the light-absorbing portion of GFP, the chromophore.

In order to get new DNA, Prasher followed in Shimomura's footsteps and went to Friday Harbor to collect more *Aequorea* tissue, which he froze in liquid nitrogen and took back with him to Woods Hole. While at Friday Harbor, he gave a talk about his research, and one of the people attending the talk was Osamu Shimomura. Although both Shimomura and Prasher had the same interests in GFP and aequorin, and they both worked at Woods Hole, this was the first and only time they met. Like most other researchers, Shimomura thought that GFP would require other *aequorean* enzymes before it would fluoresce and that Prasher's ideas of using GFP as a fluorescent protein tag were very ambitious and had a low probability of success. Nevertheless, Prasher persevered.

Back in Woods Hole with his frozen jellyfish, Prasher went back to the lab to try to find the GFP gene, but it wasn't easy. He cut up the jellyfish DNA into 1.4 million pieces and searched through all those pieces to find a piece that made GFP. Since the jellyfish makes a significant amount of GFP, one would expect about a thousand GFP genes in the 1.4 million DNA fragments. However, Prasher found only one copy of the GFP gene. Luckily, this time it contained the amino acid sequence for the complete GFP molecule.

Having found the gene, he was able to use fairly standard methods to determine the number and the order of the amino acids in GFP—in other words, to sequence GFP. The gene was made up of 238 amino acids. However, there was nothing special about the sequence that would explain why GFP is fluorescent.

Let's go back to the cookbook analogy to explain what happened. Doug paged through all of his cookbooks looking for a dessert recipe. Unfortunately, he couldn't remember the name of the dessert. However, he did remember some of the instructions he needed to make the dessert. So he started looking for the instructions (the partial sequence). When he finally found it, he discovered that the first page of the recipe was missing. He still didn't have all the instructions, nor did he know the name of the dessert. So he went to the bookshop and paged through all the cookbooks to find the instructions. It was his favorite food, and he was confident it would appear in more than one cookbook. It didn't, and he found only one recipe containing the instructions he remembered. Now that he had the instructions (gene sequence), he wanted to make a photocopy of the recipe, take it home, and make the dessert.

For the first part of his task, Prasher used bacteria as biomolecular photocopiers. Bacteria multiply very rapidly. They do this by very quickly making exact copies of their own DNA. Bacteria, which contain a ring of DNA called a *plasmid*, are commonly used. The plasmid is removed and cut open by using enzymes. Now the foreign gene (GFP) can be inserted at the position where the cut was made, and the plasmid can be closed using a different set of enzymes before it is reinserted into the bacterium. Prasher stuck DNA containing the GFP gene into a bacterium and let it multiply away. Then he harvested the GFP genes. The biomolecular techniques for doing this type of "photocopying" have been around since 1974, when scientists isolated and copied genes for the first time. This is *cloning*.

Prasher was all set. Since the GFP gene contained the instructions for the exact order of the 238 amino acids required, and all organisms read genes and make proteins in the same way, any organism should be able to make the GFP protein from its gene. Unfortunately for Prasher, the GFP that the bacteria made did not glow.[4] He would have to wait a little longer before he could get his dessert just the way he wanted it.

Prasher's GFP work was funded by the American Cancer Society. In his grant he suggested that it should be possible to take the GFP gene out of the jellyfish cell and attach it to cancer cells so that they would be labeled with a fluorescent tag. Sadly he had only a two-year grant, and the funding ran out before he could spend enough time trying to express the GFP clone he had produced. Prasher had written grants to other funding agencies such as the National Institutes of Health, but they were not interested in funding his work before he could show that GFP could be expressed in organisms other than jellyfish. Since he had funds from the Office of Naval Research to try modifying aequorin so that it would glow when it bound heavy metals like mercury, Prasher concentrated on this work instead of his GFP research. Fortunately, two other researchers, Roger Tsien and Marty Chalfie, were interested in his work, and they independently called him requesting his GFP clone so that they could try to express it. They wanted to use the GFP gene Prasher had isolated and multiplied to make GFP in some of the organisms they were studying. Prasher sent them the clones with little hope of success. He was therefore pleasantly surprised when he got a phone call from Marty Chalfie informing him that they had managed to make fluorescent bacteria. GFP was able to form its chromophore in the intestinal bacteria *E. coli* and fluoresce. In the next chapter, we will meet Marty Chalfie and get to read his GFP story.

Soon after GFP was first fluorescently expressed by Chalfie, Prasher left Woods Hole. Today, he works for the National Plant Germplasm and Biotechnology Laboratory, a branch of the United States Department of Agriculture. He no longer does research with green fluorescent protein, but like a proud father, he still keeps up with the new developments in GFP research.

Prasher's main contributions to the GFP revolution were to locate the GFP gene, sequence it, and clone it. If we once again go back to the cookbook analogy, Prasher found the GFP recipe in the cookbook, he determined the instructions (sequenced the gene), and he photocopied the recipe (cloned the gene) so that he could now distribute the recipe to other researchers. His grant money ran out before he had the opportunity to try making GFP glow in another organism, but others used the recipe (gene) that he had photocopied (cloned) to take the final step that started the GFP revolution.

CHAPTER 6

THE BIRTH OF THE GREEN FLUORESCENT PROTEIN REVOLUTION

At this point, the GFP revolution was just one experiment away from lighting up a whole new submicroscopic world. Both Prasher and Shimomura, who laid down most of the ground work, had strong bioluminescence backgrounds, but Martin Chalfie, the man who would give birth to the GFP revolution, was a biologist interested in the roundworm, *Caenorhabditis elegans*. He had never heard of GFP or *Aequorea victoria*. How, then, did Marty Chalfie become the father of the GFP revolution? What made him devote his life to studying tiny little roundworms that are barely visible to the naked eye?

Martin Chalfie was born in Chicago and grew up in Skokie, a suburb of the city. There were no scientists or engineers in Marty's family background. His father was a professional guitarist before he got married and never finished high school. His mother had to leave college before completing her undergraduate education because of the Depression. She started working in a dress-making factory that was owned by her family; later in her life, she would run it with her husband and brother working as salesmen for the company. Nevertheless, Marty was always interested in science: As a child, he cut out pictures of animals and read about astronomy. There was no particular area of science he found more interesting than any other. When he went to college, he still had no idea what area of science he was going to major in, and for a brief time while an undergraduate at Harvard, he flirted with the idea of majoring in mathematics, but finally he majored in biochemistry.

Most academics tend to go straight to graduate school after graduation, but Marty took a few years off. He did some odd jobs, including a couple of years of teaching chemistry, sociology, and mathematics at a high school. He also worked in a lab, which rekindled his interest in science and research. So he went back to Harvard University to do his doctoral work in the physiology department.[1]

After receiving his PhD, Chalfie decided to do a postdoc with Sydney Brenner and work with him on his *C. elegans* project. Doing a postdoc with Brenner meant that Chalfie had to move from Cambridge, Massachusetts, to Cambridge University, England. Chalfie chose to do a postdoc with Brenner because he was doing stimulating and important research in molecular biology. In addition, an old school friend of Chalfie's, Robert Horvitz, was doing a postdoc in the Brenner lab. While visiting Marty at Harvard, Horvitz told him all about the *C. elegans* work they were pursuing on the other side of the Atlantic. Marty thought the work sounded very exciting and applied for a postdoctoral position with Brenner.

In order to get a tenure-track position in science at a good research university, one has "to do" a postdoc. They used to be a year or two long, but postdocs of three and four years are not unusual now. Working on his postdoc with Sydney Brenner was an excellent choice, since Brenner was one of the most important researchers in modern genetics. After collaborating with Francis Crick in the early 1960s and being responsible for numerous breakthroughs in molecular biological research, Brenner had decided that it was time to move on to "other problems of biology, which are new, mysterious and exciting. Broadly speaking the fields which we should enter are development and the nervous systems."[2] In order to accomplish this, Brenner and his coworkers, including Martin Chalfie, pioneered the use of *C. elegans* as a model system. In 2002, together with John Sulston and Robert Horvitz, Sydney Brenner won the Nobel Prize in Medicine for discoveries concerning genetic regulation of organ development and programmed cell death. The prize was for work that Brenner had done over the last thirty years. Both Sulston and Horvitz had worked in Brenner's lab as students and then started their own *C. elegans* labs. The Nobel Foundation praised their achievements,

> The human body consists of hundreds of cell types, all originating from the fertilized egg. During the embryonic and fetal periods, the number of cells increase dramatically. The cells mature and become specialized to form the various tissues and organs of the body. Large numbers of cells are formed also in the adult body. In parallel with this generation of new cells, cell death is a normal process, both in the fetus and adult, to maintain the appropriate number of cells in the tissues. This delicate, controlled elimination of cells is called programmed cell death. This year's Nobel Laureates in Physiology or Medicine have made seminal discoveries concerning the genetic regulation of organ development and programmed cell death. By establishing and using the nematode *Caenorhabditis elegans* as an experimental model system, possibilities were opened to follow cell division and differentiation from the fertilized egg to the adult. The Laureates have identified key genes regulating organ development and programmed cell death and have shown that corresponding genes exist in higher species, including man. The discoveries are important for medical research and have shed new light on the pathogenesis of many diseases. Sydney Brenner (b. 1927) established *C. elegans* as a novel experimental model organism. This provided a unique opportunity to link genetic analysis to cell division, differentiation and organ development—and to follow these processes under the microscope.[3]

In 1982, Marty joined the faculty at Columbia University, and he has been there ever since. Following his postdoc with Sydney Brenner, he was hooked on *C. elegans*, and most of his research at Columbia revolves around this little barely visible worm. Chalfie and many other biologists are extremely interested in *C. elegans* because it is one of the simplest organisms that has many physiological similarities with the human body. It produces sperm and eggs; the baby worm is conceived from a single egg; it undergoes a complex development to form the adult worm; it mates; it ages; and it has a brain. It is a small roundworm, no longer than one millimeter, and lives in the soil. Since it is transparent, one can easily monitor the development of its internal organs by using a microscope. It has a relatively short life span and grows from egg to an egg-laying adult in three days, which is beneficial to studying its genetics. In 1988 the entire genome of *C. elegans* was sequenced—it was the first animal genome completed. It really is a good model for human genetics since 30 percent of its genes are related to human genes. This led Bruce Alberts, president of the National Academy of Sciences, to remark: "We have come to realize humans are more like worms than we ever imagined."[4] Of course, there are also many, many differences between *C. elegans* and humans. Most *C. elegans* are hermaphrodites, and only about 0.05 percent of normal laboratory populations are males. These males can fertilize the hermaphrodites, but since there are so few males, most fertilization occurs by self-fertilization, which has numerous advantages in genetic analysis. Most of the volume of the worm is taken up by its reproductive system.

In 1988, just as Doug Prasher had started to work on the sequencing and cloning of GFP, Martin Chalfie heard about GFP for the first time. He was attending a seminar on bioluminescent organisms at Columbia University. As part of the talk, the speaker mentioned GFP. That it was an inherently fluorescent protein got Chalfie very excited. This was the moment of conception of the GFP revolution. Chalfie immediately began envisioning ways he would be able to use GFP in his *C. elegans* research. He was so intrigued by the potential uses of the protein that he didn't listen to a word of the rest of the seminar. Chalfie wanted to use GFP as a marker that could be attached to a promoter. As noted, the promoter is a region of DNA located in front of a gene; when the cell needs to make a specific protein, it binds to the promoter for that gene, which, in turn, activates the gene. By attaching GFP to a promoter, Chalfie was hoping that GFP would be produced whenever the promoter it was attached to was activated; in this way, GFP fluorescence could be used to signal activation of the GFP-tagged promoter. After a couple of days of inquiry, Chalfie managed to find out that Douglas Prasher was working on the sequencing and cloning of GFP. He called him to share his ideas with him and to see whether Prasher would be interested in collaborating. Chalfie thought that his roundworm, *C. elegans*, would be a great organism on which to test GFP because all 959 cells of its transparent body are visible with a microscope, so any GFP made would be observable. Chalfie had also developed cell-specific promoters, which meant he had the ability to attach the GFP gene to the promoter genes in all the cells, but to activate only promoters that were located in certain cells. By adding GFP to different promoters, Chalfie was hoping to turn his roundworms into miniature flash lights, which lit up whenever its promoters were activated.

Prasher was very interested in collaborating, but at the time he had just finished sequencing the partial DNA fragment for GFP, and he realized that he had to find the complete gene and start over again. As we have seen in the last chapter, Prasher had already envisioned using GFP as a fluorescent marker, but he was much more cautious than Chalfie. Still, he promised to call Chalfie as soon as he had completed the task. The reason for Prasher's pessimism was that he knew that the chromophore was formed by the protein binding to itself and forming a small loop. But he was never convinced that the green fluorescent protein did this all by itself. Prasher thought that some other proteins were required to help GFP make its chromophore and become fluorescent. This meant that the other proteins that aided in cyclization always had to be present if one wanted to observe fluorescence. Since that was unlikely to be the case when the GFP gene was expressed in organisms other than *Aequorea victoria*, Prasher was less optimistic than Chalfie. Fortunately, he persevered with sequencing and cloning GFP, but it took a while before he managed to sequence and clone the complete GFP gene.

When Doug Prasher completed sequencing and cloning the complete GFP gene in 1992, Chalfie was at the University of Utah. Prasher called Chalfie to inform him that he was done with the sequence and that Chalfie's group should try using it as a marker for promoter activity. However, for some reason, when Prasher called the University of Utah, his call was not forwarded to Chalfie, and he was unable to get hold of him. Prasher decided to go ahead, and he published the results of his cloning and sequencing work in the journal *Gene*.[5] There are so many papers published each week that it is impossible to keep track of everything that has been released. Since Chalfie was mainly interested in *C. elegans*, he missed the paper that Prasher had written and never found out that GFP had been cloned and sequenced. Once back at Columbia University, Chalfie got a graduate student, Ghia Euskirchen, who was doing a rotation in his lab.

At Columbia and many other graduate schools, graduate students get to do a few rotations before they choose their doctoral advisers. The main aim of the rotations is for the students to get a taste of all the different research that is being conducted in their area of interest. It is a huge commitment to start one's PhD with a specific research adviser. A typical PhD in the biological sciences takes between four and eight years. Thus, it is a good idea to select an area that will interest the student for a long and intense period. It is also important for the student to find a research group and adviser with compatible goals and personalities. Rotation projects are often small, isolated research projects or an offshoot of someone else's graduate work.

Chalfie told Ghia Euskirchen all about fluorescent proteins and what he wanted to do with them. Since he hadn't heard back from Prasher, and Columbia University had just installed Medline, a database of biomedical and life sciences publications, on its network, Chalfie did a literature search together with Euskirchen. They decided to start the search by looking for GFP. Chalfie didn't think he would find anything, since he hadn't heard back from Prasher, but GFP was the only fluorescent protein he knew about, so it would be a good place to start the search. Medline contains about eleven million records from more than seventy-three hundred different publications dating from 1965 to the present and is updated weekly. One of the first hits in Euskirchen and Chalfie's search was Prasher's *Gene* paper. They ran down to the library and found the paper. Today, they could have stayed in Chalfie's office and downloaded an online version of the paper in less than a minute. The paper looked very promising: Prasher had sequenced and cloned GFP; however, at the end of the paper he refers to the sequenced GFP as *apo-GFP*, meaning that he thought the GFP that he had sequenced did not contain the chromophore. He couldn't get his cloned GFP to fluoresce and presumed this was due to the fact that other proteins were required to help with chromophore formation. Despite this, Chalfie still thought that trying to get GFP to fluoresce in *C. elegans* and then attempting to use it as a tracer molecule would be a good set of experiments for a rotation student to try. So since the manuscript had Prasher's phone number on it, Chalfie called him to see whether he was still interested in collaborating.[6]

"What's going on? Why didn't you call me?" Chalfie asked.

"I tried," was Prasher's reply, and he explained how he hadn't got through to Chalfie in Utah and that he was still interested in collaborating, although he wasn't very optimistic that Chalfie's ideas would work because he hadn't seen any fluorescence when he had cloned GFP. Nevertheless, he sent Chalfie his clone. About a month after getting instructions from Chalfie on how to incorporate the GFP gene into *E. coli*, Euskirchen succeeded. She immediately got hold of Chalfie and brought him to look at the *E. coli* through the microscope—the bacteria fluoresced green when they were irradiated with blue light! Once she showed her adviser the fluorescent bacteria, he immediately called Prasher, and together they devised some more experiments to show that the GFP gene could be expressed to produce fluorescent protein in other organisms. Having *E. coli* make GFP was a great success; it showed that there was nothing in *Aequorea* besides GFP that was required to make it fluoresce. Once the GFP gene was inserted in *E. coli*, it could easily be tricked into making green fluorescent protein that fluoresced.

Chalfie wasn't the only one who tried expressing GFP in bacteria like *E. coli*. Roger Tsien, who has been instrumental in making GFP the incredibly useful biological technique it is today, started working on GFP the same time as Chalfie. However, he didn't have a molecular biologist in the lab who could express GFP in other organisms, and he was scooped by Chalfie. Prasher himself tried, and there were probably others, but they were not able to get GFP to fluorescence. Chalfie succeeded because he asked Euskirchen to use a polymerase chain reaction (PCR) to multiply only the GFP gene. PCR is the same technique used to accurately multiply DNA in forensic fingerprinting. It worked very well for Euskirchen and Chalfie. The other researchers used restriction enzymes to cut out the GFP gene—these are proteins that cut DNA at specific positions. Unfortunately, the method left DNA bases on each side of the GFP gene, which prevented the expression of the DNA. On the other hand, there might not have been many researchers who spent much effort to try to express the GFP gene in other organisms because they thought the same as William Ward of the Center for Research and Education in Bioluminescence and Biotechnology at Rutgers University, who wrote, "Along with others, I believed formation of the GFP chromophore to require additional enzymes—thus GFP could not be cloned from a single gene. An autocatalytic reaction cyclizing three amino acids in the primary sequence had never been considered. It took a developmental biologist, Marty Chalfie, not knowing the cloning of GFP is 'impossible' to do the 'impossible.'"[7] Ward's ideas were the same as those held by most scientists; this was because most plants and animals owe their bright colors and fluorescence to the physical organization of tissues and/or the presence of pigment molecules. The wings of butterflies and the scales of tropical fish are examples of vivid colors that are due to the intricate ultrafine physical organization of tissues that cause differential scattering of incoming light.[8] The bright colors found in the tropical tree frogs are the result of small pigment molecules. The biosynthesis of these pigments is a complicated process that proceeds through several sequential reactions each catalyzed by a specific enzyme. The GFP family of proteins are the only proteins in which the pigment is not formed by an enzyme. Green fluorescent proteins catalyze the production of their own pigment, and since it remains part of the protein, it is called a chromophore. Like most pigments, the chromophore of GFP absorbs light because it has some double bonds alternating with single bonds. Systems with many alternating single and double bonds are called *conjugated systems*, and they are very effective at absorbing light. Diamond and charcoal have exactly the same atomic composition: both are made up of nothing else but carbon atoms. In a diamond there are only single bonds between the carbons, and it absorbs no light. In charcoal there are alternating single and double bonds creating extensive conjugation, which is why charcoal absorbs visible light and is black. The chromophore of GFP is somewhere between diamond and charcoal in its conjugation.

Additional experiments were now needed to show that GFP was not toxic, that it could be made, that it fluoresced in more complex organisms than *E. coli*, and that a promoter could be used to start GFP expression. Chalfie, Prasher, Ward, and one of Chalfie's technicians managed to transform *C. elegans* so that it produced fluorescent GFP. The addition of GFP to a promoter that activated genes only in the neurons of *C. elegans* resulted in GFP fluorescence that was localized to the neurons. Besides fluorescing under UV light, the transformed *C. elegans* developed and functioned no differently from normal *C. elegans*, which showed that GFP was nontoxic and did not impair the function of other proteins in the roundworm.[9] However, it wasn't all smooth sailing. Chalfie's lab wasn't set up to do research on fluorescence, and the fluorescence microscope they had was inadequate for seeing the fluorescence in the tiny *C. elegans*. So the technician working on the project had to scrounge for microscopes. Since this was a new area of research for Chalfie and it was developing so fast, he never had the opportunity to write a grant to fund his GFP work. Writing grants takes some effort and time, and once a grant has been submitted, it is very rare to get funding in less than six months.

In 1992, while these initial experiments were still being conducted and before GFP was expressed in *C. elegans*, Chalfie went to a neurobiology conference where in confidence he told a number of people about the fluorescent protein he was using as a marker. One of the people he told about GFP was Roger Tsien, who, in the coming years, would be instrumental in modifying GFP and in finding many new uses for it.

The GFP revolution had begun. When they were ready, Chalfie and his coworkers published their results in the February 11, 1994, issue of *Science*.[10] There are hundreds of different science journals, and there is an art to publishing in the right journals and being able to distinguish the quality of the different journals. Most are peer reviewed, which means that an author submits his research to the editor of the journal, who, in turn, sends the paper to three or more external referees. These scientists then have to decide whether the work is worth publishing in the journal. *Science* is the Ferrari of the scientific publishing world. Only the best papers that will be of importance to a wide range of scientists get published there.

In their *Science* paper, Chalfie and his coworkers wrote:

> Several methods are available to monitor gene activity and protein distribution within cells. These include the formation of fusion proteins with coding sequences for [beta]-galactosidase, firefly luciferase, and bacterial luciferase. Because such methods require exogenously added substrates or co-factors, they are of limited use with living tissue. Because the detection of intracellular GFP requires only irradiation by near UV or blue light, it is not limited by the availability of substrates. Thus, it should provide an excellent means for monitoring gene expression and protein localization in living cells. Because it does not appear to interfere with cell growth and function, GFP should also be a convenient indicator of transformation and one that could allow cells to be separated with fluorescence-activated cell sorting.[11]

Basically, he was saying that GFP was extremely useful since it fluoresces without the addition of any other chemicals; luciferase like all the other known methods required the injection of an additional substance such as luciferin before they could produce light.

A very important experiment that Chalfie and his coworkers did not do was to tag a protein with GFP. The first person to do this was Tulle Hazelrigg, the wife of Marty Chalfie. Perhaps she should be the mother of the GFP revolution. She had just moved to Columbia University to be with her husband, Marty, and was setting up her lab. One of her first projects at Columbia was to fuse GFP to a protein that was found in the fruit fly. This was done by placing the gene for GFP at the end of the gene of interest in the DNA of the fruit fly. Using GFP fluorescence, Hazelrigg was able to show that the protein was initially localized to one part of the cell and that it was subsequently distributed in a unique way that she could follow over a period of time.[12] She published her results in *Nature*, which is the British equivalent of *Science*. There is no better place to publish a paper than in *Nature* or *Science*. Marty says Tulle might have got a step ahead of her competitors because she was at Columbia University, where she heard a lot about GFP from the students, and because she could hardly escape his enthusiasm about GFP. But she wasn't the only one; everyone in the Columbia biology department, and everyone who met Marty in these exciting months, got to hear about GFP and its potential uses. Tulle used GFP as a protein tag because she was able to see the promise the technique had. Marty certainly never asked her to use GFP in her research, and the fact that they were married did not help her much.[13]

While Marty uses *C. elegans* in all his research, Tulle uses *Drosophila*, the fruit fly, as her model organism. Fruit flies have been used to study genetics for more than one hundred years. They are also known as vinegar flies, garbage flies, and banana flies. They are the tiny, delicate flies that often buzz around decaying fruits. Thomas Hunt Morgan in 1904 chose to study the hereditary traits of the common fruit fly, *Drosophila melanogaster*. He too did all his work at Columbia University. He chose the fruit fly for his studies because they breed much faster than rabbits; millions can be bred in simple milk bottles, and they cost nearly nothing to house, grow, and feed. Together with research students, Morgan crossbred millions of flies and examined them under a microscope to see if they could find some genetic variations. Despite zapping their flies with radiation, spinning them in centrifuges, and gently heating them, they could find no genetic mutations. After six years, Morgan was about to give up when he found some fruit flies with white eyes in place of the usual red eyes, a hereditary characteristic that he could monitor. It was all he needed. In 1933 he was awarded the Nobel Prize in Medicine for his discoveries concerning the role played by the chromosome in heredity.[14] Since that time, the entire genome of the fruit fly has been sequenced, and another set of Nobel Prizes in Medicine has been awarded to fruit fly researchers. Fruit flies are used by thousands of scientists throughout the world, so showing that you could tag proteins and observe when they are produced and what their movements are in a live fruit fly was a very great breakthrough.

Despite being used as a fluorescent marker for promoters in *C. elegans* and as a tag for proteins in the fruit fly, the remaining scientific community took some time before appreciating the usefulness of GFP and using it in their research. This surprised Marty. Interestingly enough, a large number of the early requests for the GFP clones came from researchers who had heard about the usefulness of the protein as a marker from their students, who, in turn, had heard about it from the graduate research student grapevine.

As time went by, Marty got a number of phone calls from more and more researchers interested in using GFP in their research. One of their first questions was always whether GFP had been used in their organism before. Usually the answer was no. This didn't produce the response he expected. Many callers did not want to pioneer the use of GFP in new organisms or cell types; they were content to wait for others to find and solve all the problems. He had expected them to be excited about the possibility of being the first to use GFP and get all the associated citations and accolades.[15] However, that was not the case; perhaps this might be due to molecular biologists getting most of their chemical reagents in kits that are as simple to use as it is to set up an Apple Mac. The attitude of biotech companies also surprised Marty. At first he had thought that they would see the potential of GFP markers and develop kits. However, it was the other way around. The biotech companies waited for researchers in the academic institutions to optimize the GFP reporters before they converted them into commercially available kits.

While Shimomura can be thought of as a reluctant grandfather of the GFP revolution, Marty Chalfie is the enthusiastic and proud father of this revolution. He smiles and laughs when he talks about GFP and all its uses. He doesn't see the need to compare his two children; that is, his *C. elegans* work is completely different from his foray into GFP chemistry.

Although he had a very important role to play in developing GFP as a marker molecule, Martin Chalfie never seriously pursued GFP research—until very recently, that is. At first he made a half-hearted attempt to carry on with the GFP research, but that gradually fizzled out. Only one person ever contacted Chalfie with a serious inquiry to do a GFP postdoc. He wanted to put GFP into plants and was under the false impression that the only way he could get to do this experiment was by doing a postdoc in Chalfie's lab. When Marty offered to send him the GFP clone instead, he lost the only postdoc who ever applied to do GFP research in his research group. He doesn't seem concerned and gets more than his fair share of students who are interested in working on *C. elegans* in his lab. Marty still follows the latest developments in GFP technology. He is editing the second edition of a scholarly book about GFP and protocols for its applications and has just published a paper describing new uses for modified GFP (see chapter 9).[16]

One day not long after Martin Chalfie had published his paper about using GFP as a tracer molecule, he, Tulle, and their daughter went shopping and went into a pet store. There they saw a wonderful display of fish swimming under a fluorescent light. All the fish had amazingly different glowing colors. The sign next to the fish tank said "painted glassfish," a species of fish neither Tulle nor Marty had heard of before. An excited Chalfie went to the store proprietor and asked him about the "painted glassfish." Did they contain a fluorescent protein he did not know about? The man told him to calm down—if he wanted to buy the fish he should feel free to do that, but he should be aware that the paint would come off in a month or so.

Marty Chalfie was intrigued and asked the shopkeeper if he thought that there would be any commercial interest in genetically modified fluorescent fish. "Sure thing," replied the proprietor. "You would sell millions." Little did Chalfie know that in less than ten years, transgenic GFP zebra fish would indeed be sold as pets in Taiwan and the United States and that a green fluorescent GFP bunny would cause a media frenzy.

GFP research didn't end after Chalfie showed that it could be expressed in any organism, that it could be used as a reporter molecule, and that it was safe. Doug Prasher collaborated with some other researchers to establish why GFP could fluoresce without the addition of any other chemicals. What they found was that when GFP is made, it folds in such away that the 67th amino acid attacks the 65th amino acid to form the structure predicted by Shimomura. This is truly remarkable. There are tens of thousands of proteins, each with hundreds of amino acids, and none of these amino acids attack each other. Yet in GFP they do, and they form the chromophore that gives off the green fluorescence that is observed in the jellyfish. Prasher and his coworkers were able to determine how the chromophore is formed, but they didn't know why it formed in GFP and not in any other proteins. They answered an important question, but in doing that found that there was another layer of questions that still needed to be answered. This is a very common occurrence in science.

I first heard of GFP in 1995 when Doug Prasher gave a seminar about his role in sequencing the GFP gene at Connecticut College. I was amazed to hear that once GFP had been expressed, the 238 amino acids folded in such a way that one of the amino acids attacked the carbon of an amino acid located two amino acids away from it in order to form the chromophore. This occurred in every organism GFP had been expressed in, which meant that no other chemicals were required. There are more than one hundred thousand proteins, yet GFP is the only protein that attacks itself. Why? This question has fascinated me for the last eight years, and my computers have spent thousands of hours calculating different mechanisms of chromophore formation. One of the things we have found is that the amino acids in the chromophore-forming region are in an unusually tight loop. This holds the two amino acids that join chromophore in proximity to each other. However, there is a lot more to it than that. I find my calculations fascinating, but I am most likely in the minority, and they probably don't belong in a book like this. Rather than take the risk of boring you, I think it's time to have a look at some of the amazing applications for green fluorescent protein that have appeared in the scientific literature over the last ten years.

CHAPTER 7

THIRSTY POTATOES AND GREEN BLOOD

In 1674 Anton van Leeuwenhoek invented the simple microscope, and it opened up a whole new world. Suddenly, one could see things that had never been visible before. His microscope was able to magnify objects up to three hundred times, and he was one of the first people to observe spermatozoa, blood cells, bacteria, and muscle fibers. On September 17, 1683, Leeuwenhoek wrote to the Royal Society about the plaque between his own teeth. This is the first reported observation of bacteria, which he called *animalcules*: "a little white matter, which is as thick as if 'twere batter. . . . I then most always saw, with great wonder, that in the said matter there were many very little living animalcules, very prettily a-moving. The biggest sort . . . had a very strong and swift motion, and shot through the water [or spittle] like a pike does through the water. The second sort . . . oft-times spun round like a top . . . and these were far more in number."[1]

GFP has become the microscope of the twenty-first century. Every month more than two hundred papers are published reporting yet another way GFP has been put to work. In most cases, GFP can be used in a way very similar to a microscope; it can show us when a protein is made and what its movements are. Still, this is not always the case. Sometimes GFP has been used solely for the allure of its fluorescence. Recall Marty Chalfie's excitement when he saw fluorescent fish in a local pet shop. His euphoria didn't last long because he soon discovered that they had been painted with fluorescent paint.

In Taiwan it is possible to buy transgenic GFP zebra fish that really are fluorescent. (See figure 4 in the photo insert.) They are called *night pearls* and are sold for about $15 apiece. *Time* magazine named them one of the coolest inventions of 2003. H. J. Tsai, a microbiologist at the National University of Taiwan, originally created them to study organ development in zebra fish; initially he wanted to label just the zebra fish organs with GFP. However, much to his surprise, the whole zebra fish glowed due to GFP expression.[2] Tsai took some pictures of the fluorescent fish and forgot about them until he decided to show the pictures at a conference. A representative from the Taikong company, which sells aquarium equipment and fish food all over the world, saw the slide and agreed with Tsai to fund his research. They invested $2.9 million to get to the point where they can now produce transgenic GFP zebra fish en masse for the pet trade in fisheries located outside of Taipei. They are sterile and are sold with a UV lamp, plastic fluorescent coral, and fluorescent fish food. Zebra fish with both green fluorescent proteins from the jellyfish and red fluorescent corals have been created. By May 2003, Taikong had more than $2.3 million of foreign orders for its glowing fish. It hasn't been easy everywhere—some countries like Singapore have banned the sale of transgenic pets and have confiscated four hundred night pearls. In Japan, a law that requires a certification that the genetically modified fish is safe was introduced in January 2004. Tsai and the Taikong Corporation are currently trying to create glowing dragon fish. They chose the dragon fish because it is considered a lucky species by many Asian cultures.

A group of researchers from Singapore University has gone one step further. They have created a series of zebra fish in a whole spectrum of colors. Green fluorescent protein (GFP), yellow fluorescent protein (YFP), and red fluorescent protein (DsRed) were expressed in the skeletal muscle of zebra fish by attaching their genes to a strong muscle-specific promoter, *mylz2*.[3] The resultant transgenic zebra fish have bright green, red, yellow, and orange fluorescent colors that are visible to the naked eye under daylight and ultraviolet light. The orange fish were created by expressing a combination of DsRed and GFP in the muscle. These fish, created by Zhiyuan Gong's group, are not sterile and can reproduce; however, they are no more efficient at surviving or reproducing than other zebra fish. This means that they could probably safely be used in the ornamental fish market, as most ornamental fish species cannot survive in the wild after having undergone generations of breeding and selection in the controlled aquaria environments. An aquarium of fish fluorescing in many colors will attract many ornamental fish fans. Gong's fish, however, are most interesting because between 3 and 17 percent of the muscle protein in the transgenic fish is made up of fluorescent proteins. The muscle-specific mylz2 promoter could therefore be used to convert zebra fish into protein factories; each gram of wet-muscle tissue could generate between 4.8 and 27.2 milligrams of a desired protein. This is comparable to or even better than the levels of recombinant proteins expressed in the mammary gland systems of transgenic farm animals such as pigs. Since there are estimates that more than 20 percent of all pharmaceuticals will be produced by transgenic organisms by the year 2020, one can perhaps consider these transgenic ornamental fish the precursors to a new generation of swimming pharmaceutical factories. To create even more successful and efficient bioreactor systems, Gong and his coworkers are currently researching the use of fast-growing and large farm fish such as carp, tilapia, catfish, salmon, and rainbow trout.

One of the problems with writing a book like this and doing research in the green fluorescent protein field is that it is growing so rapidly. A month or so after I had completed my first draft of this chapter, there was a flurry of articles about fluorescent ornamental fish in the American press. A Texas company, GloFish, had announced that it was going to sell fluorescent genetically modified zebra fish starting in January 2004. The company was using Gong's technology to produce red fluorescent zebra fish with the DsRed gene; in turn, he would get a portion of all the proceeds from the sale of all GloFish fluorescent fish in America. GloFish was started by Alan Blake, twenty-six, and a partner in Austin, Texas, two years ago. The fish were bred and distributed by two tropical fish wholesalers, Segrest Farms in Gibsonton, Florida, and 5-D Tropical in Plant City, Florida. The only other fish with colors as vibrant as the fluorescent fish are saltwater fish, which are much more difficult to maintain in an aquarium. GloFish can currently be bought in forty-nine of the American states. Only California has banned the sale of transgenically modified pets. Many different groups nonetheless have expressed some concern about the sale of genetically engineered pets. Since GloFish are not meant to be eaten, neither the FDA, EPA, nor USDA have the will or power to regulate them.

Les MacPherson of the *Star Phoenix* (Saskatoon, Saskatchewan) has written a very amusing article describing the GloFish and summing up some of the fears people have about the creation of genetically modified pets. Describing the fish's glow, he writes, "At least for the moment, GloFish are no threat to the powerful light bulb industry. Strictly speaking, GloFish don't really generate any light at all. They only seem to glow when exposed to ultra-violet light, rather like the old Jimi Hendrix poster that's been rolled up on the top shelf of your closet ever since you got married. Now, thanks to the miracle of genetic engineering, the same 1960s, glow-in-the-dark technology that's not good enough anymore to have up on the wall in your basement is available in a fish."[4] Although the caution he expresses in the next excerpt of his article is pretty frivolous, I think we all get the idea: "This is not to suggest that caution be thrown to the wind. Genetically modified organisms really are a potential threat to other species. Scientists who do this kind of work must be very careful. We don't need venomous budgies, for example. We don't need flesh-eating hamsters. We don't need flying pit bulls. Something like a flying schnauzer, however, might be fun. The only species threatened by a flying schnauzer would be cats."[5]

The consensus of opinion among scientists seems to be that if GloFish are released into the wild, they will not be a danger to the native fish populations. Frank Greco, a senior aquarist at the New York Aquarium, said, "[The GloFish] are perfect targets. It's like they have a neon sign on them that says 'Eat me.'"[6] Purdue University geneticist William Muir is not worried about GloFish being released into the environment, however. He points out it is impossible to prove a negative and to show that GloFish won't have some effect on the environment. Muir is much more worried that some other less scrupulous company will produce other types of transgenic fish such as fluorescent colored carp, which will sell very well, get loose, and cause havoc in the environment.

Zebra fish are transparent, making it very easy to see GFP's fluorescence. Creating fluorescent flowers is not as easy. The molecular basis of coloring in flowers is extremely complex and is often due to the expression of numerous different floral pigments, many of which can absorb or reflect ultraviolet radiation and thereby interfere with GFP's fluorescence. Since white flowers have fewer color pigments to interfere with GFP's glow, Tito Schiva of the Experimental Institute of Floriculture, San Remo, Italy, used white daisies in an experiment to demonstrate that white flowers could be made fluorescent by introducing a GFP gene. It is unlikely that fluorescent flowers will become commercially available, especially in Italy, given that genetically modified organisms are very unpopular in Europe and that to get governmental approval to sell them would cost at least $1 million. Figure 5 in the photo insert shows a transgenic GFP daisy next to a normal daisy or osteospermum, under ultraviolet light.[7] Schiva and his coworkers were originally trying to create transgenic GFP marijuana plants, so that authorities could distinguish between legitimate fluorescent hemp crops and narcotics, but he had trouble observing the fluorescence in the marijuana plants. That's why he tried expressing GFP in white flowers.

Creating fluorescent plants is more than just a clever laboratory trick. The technique has some very down-to-earth applications. For example, I never know how much and when to water flowers left in my care, which probably is why I am no longer trusted with plants. Thanks to *Aequorea*, that may change one day. Researchers at Edinburgh University have developed transgenic GFP potatoes that have leaves that start fluorescing green when they need to be watered. They were created by finding a protein that was produced in the potato leaves whenever the plant was not getting enough water and tagging the GFP gene to the gene for the protein. The work was done by Anthony Trewavas, whom we have met in chapter 4, where I described how he used aequorin to monitor calcium concentrations in tobacco plants. The potatoes with fluorescent leaves are not meant to be eaten but to serve as sentinels, warning the farmer when his commercial crop needs to be watered. This might sound like a very limited application, but it can save millions of dollars in irrigation costs, especially if the technique can be extended to other crops, such as corn and cotton. Sandy Bain, chair of the Scottish National Farmers' Union, said: "This kind of technology could be exceptionally valuable in determining precisely when we have to irrigate and feed our plants so we can try and ensure we get the best possible return."[8] Potatoes are an extremely important food resource; they are one of five major crops that feed three-quarters of the world's population. Farmers favor overwatering potato crops, since the plant, which originated as a crop more than six thousand years ago in the moist environs of the Andes, tends to grow larger spuds when offered an abundance of water. This is no problem if there is sufficient water; however, this might not always be the case: "There have been signals that by 2050, water could be the most expensive agricultural product in the world," said Trewavas.[9]

A number of similar GFP-based sensors have been described in the literature, and it is just a question of time before they are used in everyday situations. All the GFP sensors work in an analogous fashion and rely on a GFP-tagged promoter gene that is expressed only under certain conditions. In some cases, the promoter gene is kept in its original host; in others it is transferred into a different organism. The next example describes the use of a mouse promoter gene that has been tagged with the GFP gene and transferred from the mouse to the tadpole.

Transgenic tadpoles have been produced that give off green fluorescence in the presence of cadmium and zinc, two heavy-metal environmental pollutants. Tadpoles like these could be used in the near future to monitor pollution levels in streams, ponds, and lakes. Fortunately, extreme cadmium poisoning is extremely rare. The most well-known case occurred in the Jinzu valley in Japan, where a zinc mining and smelting company had released large quantities of cadmium into a local river. Water from this river was used to irrigate rice paddies, which led to high cadmium concentrations in the rice. Cadmium and calcium are very similar; consequently, the bone-making mechanisms in the body cannot distinguish between the two. Hundreds of people whose diet consisted of the contaminated rice developed a painful degenerative bone disease called *itai-itai*, which means "ouch-ouch" in Japanese. Old women were particularly susceptible to the disease, which made their bones porous and led to their collapse. Lower levels of cadmium poisoning are much more common than that described above. They lead to immune system suppression and lung, heart, and liver disease.[10]

In order to develop the metal ion–responsive tadpoles, the promoter for mouse metallothionein-1 was tagged with GFP. Metallothionein is a protein rich with sulfur atoms that protects the mouse from heavy-metal poisoning by binding heavy metals such as cadmium and mercury. Humans also have a metallothionein gene, which is activated in the presence of high cadmium concentrations; it protects us from cadmium until its capacity to bind cadmium is exceeded. In mice it is produced whenever the mouse is exposed to high heavy-metal concentrations. Since mice can't swim, the promoter for mouse metallothionein was tagged with GFP and placed in frog sperm nuclei, which were then microinjected into unfertilized frog eggs. The mouse metallothionein promoter in the transgenic tadpoles was activated at concentrations of cadmium and zinc as low as 0.0000005 grams per liter. It tried to turn on metallothionein production but instead found the GFP gene and started making GFP instead. At any concentrations higher than 0.0000005 grams per liter, the green fluorescence of the tadpoles increased dramatically.[11] Given the importance of clean water worldwide, the implications of having tadpoles that can signal poisoned water are tremendous. By using similar techniques with other promoter-gene systems, the researchers responsible for developing the heavy-metal sensing tadpoles hope to create more transgenic tadpoles that can serve as sensitive in vivo aquatic test systems. They could detect other toxicants, for example, sensors for endocrine-disrupting chemicals that interfere with the action of thyroid hormones.

GFP can also be used to act as a transgenic policeman, ensuring that modified genes do not cross over into other species, forming new, unwanted transgenic organisms.[12] Transgenic crops are becoming increasingly common. During 1998 50.7 million acres were planted with transgenic crops in the United States, and 68.7 million acres were used to grow genetically modified crops worldwide. People derive more than half their calories from just three cereals, wheat, rice, and maize. These cereals are rarely genetically modified, as most of the desirable traits can be obtained by traditional cross-breeding. In the United States, there is no requirement to label transgenic food products in stores.

Besides genetically modified plants and transgenic bacteria that produce bovine growth hormone, there are many agricultural products that have been genetically modified. Potatoes with chicken genes and tomatoes with flounder genes have been produced.[13] The chicken gene in potatoes produces a protein that prevents fungi from growing on the potatoes and lengthens their shelf life, while the flounder gene codes for an antifreeze gene that protects the tomatoes from cold temperatures. There is some concern that the inserted genes might wander from the transgenic crops to their surrounding wild relatives; in other words, neighboring weeds could also be made disease-, herbicide- or drought-resistant by migrating genes. To control this, beneficial genes of some of the transgenic crops can be tagged with the GFP gene. If irradiation with ultraviolet light results in green fluorescence in any neighboring plants, then we can see when and where the GFP-tagged genes have escaped their transgenically modified hosts and have entered a different species.

There are also fears that the pollen of transgenic crops might fertilize unmodified crops or have unknown side effects on ecologically vital insects. It is currently very difficult to track the movement of pollen, to study pollination mechanisms, and to differentiate pollen from individual plants of the same species. GFP-labeled tobacco pollen has been developed that can be visualized and tracked in nature, even on the legs of bees. The authors of the study hope that "[t]he expression of GFP in the pollen of plants will enable scientists to track the movement of pollen, to differentiate between pollen from individual plants of the same species, to determine pollination mechanisms, and to study spatial patterns of pollen with respect to a plant's location in the field."[14]

Prolume is an interesting biotechnology company whose core business is based on glowing genes from deep-water marine bioluminescent organisms. Initially the company concentrated on applying the glowing-gene technology to biomedical research and drug discovery. However, the company recently developed some glow-in-the-dark toys. This side business seems to be so profitable that it has spun off a new company, Biotoy. Its big seller is a $4.99 squirt gun loaded with powdered aequorin from a sea pansy. Add some distilled water to the chamber and fire away. As the liquid squirts out, it looks just like water. But when it hits something—anything that contains calcium, which can be found on people and all kinds of other things—it lights up. For $9.99 BioToy also has H_2O Glow, a bioluminescent chemistry set that allows kids to create bright blue glowing tap water using luciferase and luciferin.

Marty Chalfie was the first person to attach GFP to a promoter gene and demonstrate that he could observe fluorescence when the promoter was activated. Tulle Hazelrigg, his wife, was the first person to tag a protein with GFP. A combination of both methods was used by researchers from California to visualize how an organism smells.[15] Our sense of smell can distinguish between very subtle differences in molecular structure. For example, there are two identical forms of the molecule called *carvone*; they are made up of exactly the same atoms combined in the same way. The only difference between them is that one is left-handed and the other right-handed, yet they smell completely different—one smells like spearmint and the other like caraway. How can our nose distinguish between such tiny structural changes? One theory about smell is that our nostrils are lined with numerous receptor proteins that bind to odorant molecules. A receptor molecule is a protein that has a docking area with a very complicated shape and will bind molecules with only the exact complementary shape to the docking bay. When a molecule binds to an olfactory receptor molecule, it sends a message along a distinct neural pathway, which is ultimately interpreted as a specific smell by the brain. The perceived smell depends on which neural pathway is used. Each receptor is responsible for a slightly different smell. Humans have 347 intact olfactory receptor genes, as well as many old sensors that no longer function, probably due to disuse. The mouse has 1,036 different receptors in its nose. It is probably able to distinguish smells three times as well as we are. The 2004 Nobel Prize in Medicine was awared to Linda Buck and Richard Axel for their discoveries of odorant receptors and the organization of the olfactory system.

Adult roundworms (*C. elegans*) are often used to study smell because they have 302 neurons in their nervous system and a limited number of olfactory receptors, the identities of which are mostly known. AWA and AWC are chemosensory neurons that detect airborne attractants. In simple terms, they are parts of the brain that are activated by chemical attractants. The AWC olfactory neurons detect benzaldehyde, butanone, and isoamyl alcohol, while AWA reacts to diacetyl, pyrazine, and thiazoles. If a roundworm is exposed to an excess of diacetyl, its AWA olfactory neurons are swamped and the roundworm will not respond to pyrazine, which also binds to the AWA receptor. However, it will respond to low concentrations of benzaldehyde because the AWC neurons are not saturated by the diacetyl. When a typical population of *C. elegans* is exposed to diacetyl,

90 percent of the roundworms will accumulate at the source of the diacetyl. Careful experiments have shown that the 10 percent of round-worms that do not respond to the diacetyl molecules have a mutation in a gene called *odr-10*. Presumably the protein encoded in this gene is the olfactory receptor for diacetyl. It is deficient in 10 percent of *C. elegans*, which would explain why they don't respond to diacetyl. In order to see what the protein encoded by *odr-10* does, to find out where it goes, and to determine when it is made, two separate GFP experiments were per-formed. In the first, the *odr-10* promoter was used to drive the expression of GFP. The fluorescence associated with GFP was found only in the AWA neurons, which have cilia extending to the tip of the worm's nose. This means that the *odr-10* gene is expressed only in the AWA neurons.

In the second experiment, the GFP gene was attached to the end of the *odr-10* gene. Not surprisingly, the green fluorescence was observed in exactly the same locations as in the first experiment; however, the modified *C. elegans* fluoresced only when it was exposed to diacetyl. (See figure 5 in the photo insert.) Thus, the *odr-10* promoter is turned on and the protein is made only when the roundworm is in the presence of diacetyl.[16]

Zebra fish have been used as model systems because—like *C. ele-gans*—they are transparent and any GFP expression is easily visible. They are particularly useful in studying blood cell development because zebra fish embryos, which have virtually no circulating blood, can sur-vive for several days. This enables scientists to study mutants that have defective blood-producing systems. It is very difficult to study blood-pro-duction in mammals. Vertebrate red blood cell production is a complex process. It occurs in distinct phases and in a variety of anatomical sites. Since molecular and cellular processes associated with making red blood cells are conserved throughout vertebrate evolution, simpler vertebrates can be useful mammalian models. In zebra fish GFP has been linked to the GATA-1 promoter sequence to determine when and where red blood cells are formed in the zebra fish embryo. The GATA-1 promoter is responsible for red blood cell production. In two-day-old embryos, green fluorescence was observed in the heart. However, as the zebra fish grew into adulthood, blood production and fluorescence moved to the kid-neys.[17] These results, that blood is first produced in the heart and then later in the kidneys, have been confirmed by other techniques.

An interesting additional finding of this research was that the GFP-modified promoter remained active in future generations, and the children and grandchildren of the original transgenic zebra fish had green fluorescent blood. This was the first time that functioning genes had been permanently added to the zebra fish.

C. elegans and zebra fish are very useful model organisms because they are see-through. However, they are a lot simpler than mammals. Researchers at AntiCancer Inc. have developed a system to photograph GFP expression in the major organs of intact, live mice in a way that is affordable, simple, and fast. Using this system, scientists will be able to see when and where genes are expressed in mice without killing and dissecting the experimental animals to measure the distribution of the reporter gene.

Usually optical methods are not used to visualize and image tumor cells because tumor cells are not clearly distinguishable from normal tissue. This is not a problem when the tumor cells have been modified to express high levels of GFP. Commercial transgenic GFP human and rodent cells have been injected into mice to study how the tumor cells multiply and metastasize. Using whole-body imaging, the tumors can be observed in real time without having to invade the body of the mouse.[18] The ramifications of this can be enormous. In chapter 13, I will be telling you a lot more about how glowing genes have been used in the fight against cancer.

AntiCancer Inc. researchers have also been using GFP adenoviruses, described in the next paragraph, to examine genetic modification of mouse hair follicles. Hair grows in three phases, which are each regulated by specific molecular signals. If we could find and activate the genes that regulate the growth phase of hair follicles, then the toupee manufacturers would be out of business. However, prior to 2002, all attempts to genetically modify hair follicles had very limited success, mainly because it was very difficult to penetrate through to the hair follicle nucleus, where the DNA was stored. This problem has since been circumvented by treating skin fragments with the enzyme collagenase, which made the hair follicles accessible to an adenovirus containing the GFP gene.

Viruses are little more than bundles of genetic material surrounded by protein coats. When they come into contact with a host cell, they are able to insert their genetic material into the host. The infected cell then produces viral protein and the virus's genetic material instead of its usual products. In this experiment, a modified adenovirus was used to insert the GFP gene into the hair follicles. The skin fragments could then be grafted onto nude mice, where the GFP was expressed in as many as 75 percent of the hair follicles.[19] The experiment showed that it is possible to make efficient genetic modifications of the hair shaft, which is a major step toward preventing baldness. It may also take us a step closer to permanently changing hair color by altering the amount and type of melanin produced in the hair. Red and brown hair colors are caused by pheomelanin, and black hair results from the darker eumelanin, so by changing the genes for these melanins in the hair shaft, it is theoretically possible to "naturally" change one's hair color. (The protein that causes blonde hair is currently unknown, and so for now peroxide will have to do.) The work was initiated to aid chemotherapy patients who lost hair following treatments.

Adenoviruses are often used as molecular syringes to inject the GFP gene into specific organelles. Great care has to be taken, as the viruses are not easily controlled and one could insert the GFP gene into other undesired nuclei. Thanks to stringent FDA regulations, it is unlikely that tattoos and piercings will be replaced by a new fad—fluorescent hair, produced by using adenoviruses to insert GFP into the hair that would look normal during the day but fluoresces at night under the ultraviolet light of the rave floor.

On another note, there are some people who spend their lives studying sausages and sausage production. Mansel Griffiths at Canadian Research Institute for Food Safety is one of these people. Then there are others, and I am one of those who likes to eat an occasional sausage. My all-time favorite is the currywurst, which is a bratwurst cut into sections, smothered with tomato sauce, and then sprinkled with curry powder and paprika. Books have been written about the currywurst, and there is even a song by Herbert Groenemeyer about this delicacy from Berlin. I love sausages—the German currywurst, the American hot dog, and the South African *boerewors*—but I would rather not know how they are made and how tasty bacteria find them. Foodborne illnesses are a serious problem worldwide. Bacteria are just as fond of meat as we are. A Council for Agricultural Science and Technology report estimates that pathonogenic bacteria may cause as many as nine thousand deaths a year and between 6.5 and 33 million cases of diarrheal disease. In 1994, an *E. coli* outbreak associated with dry fermented sausage occurred in the Pacific Northwest. Dry sausage production relies on the acid produced in the process and the drying procedure to kill the pathogens; typically no heat is produced. Mansel Griffiths and his coworkers have added genetically modified bacteria that are bioluminescent not to a sausage but to a brain-heart infusion broth. Sounds gross, doesn't it! The broth was used as a model for American-style and European-style sausage production. In the American-style, a high temperature (37°C; 98.7°F) and a short fermentation time (one day) was used, while in the model for the European-style sausage production, the brain-heart infusion broth was fermented at 22°C (71.6°F) for three days. After fermentation, the samples were stored at 10°C (50°F), and the survival of the *E. coli* was monitored by using its bioluminescent properties over an extended period of time. The European-style sausage production without addition of some nitrite was the best way of making sausages—no *E. coli* survived longer than nine days. In the American-style with nitrite in the starting broth, the bacteria survived for more than fifty-one days.[20]

An alternative to nitrates as food preservatives, bacteriocins are being used; these are small toxic proteins made by bacteria in order to kill other proteins. Nisin is the only bacteriocin approved as a food preservative in more than fifty countries, including the European Union and the United States. It has been shown to have the same human toxicity as salt and is therefore commonly used. In China and Brazil, nisin is added to meats in order to prolong their shelf life. However, in the United States and the European Union, it has not been approved for the use in meat but has been used in many other food products. Recently a paper appeared in the aptly named journal *Meat Science* that describes a GFP-based method for detecting the presence of nisin in meat products.[21] The amount of fluorescence emitted by GFP in the nisin assay depends on the nisin concentration and can be used to detect nisin at concentrations of less than one molecule of nisin per million other molecules. The experiment, or *bioassay*, was used to examine the shelf life of nisin in cooked sausage. It turns out that nisin tolerates heating very well—91 percent of nisin remains alive and active after sausages have been cooked. After storing the cooked sausages in a refrigerator at 6°C (42.8°F) for twenty-eight days, 68 percent of the nisin remained active. The authors concluded that nisin has a long shelf life and should be used in conjunction with nitrates in meat products; in this way, the amount of nitrates, which are unhealthy, can be decreased without sacrificing any shelf life.

Malaria is another disease being by studied using GFP. Each year, malaria is responsible for more than 300 million acute illnesses and at least one million deaths. It is found throughout the tropical and subtropical regions of the world. Ninety percent of deaths occur in Africa, south of the Sahara, mostly among young children. Malaria is caused by a one-cell parasite called *plasmodium*, which is transmitted from person to person through the bite of a female *Anopheles* mosquito, which needs blood to feed her eggs. The parasite enters the human host when an infected mosquito sucks the person's blood. It undergoes complex life-cycle changes in the body and infects the liver and red blood cells of the host.

GFP has proved very useful in malaria research. It is extremely difficult to create transgenic mosquitoes. The first foreign gene ever placed in *Anopheles stephani*, the mosquito that transmits malaria in India, and in *Anopheles gambiae*, the most common malaria carrier in Africa, was GFP.[22] It was used because it is very easy to see when the GFP has been incorporated into the mosquito. Thanks to this research, a gene for a protein that prevents the plasmodium parasite from living in *Anopheles stephani*'s gut has been incorporated into the Indian mosquito using the same techniques as were used with GFP. Perhaps these mosquitoes will be released one day so that they can compete with the parasite-bearing wild mosquitoes.

In order to follow the lifecycle of a plasmodium parasite in living mosquitoes and mice, a plasmodium-expressing GFP was generated. To follow the GFP fluorescence emanating from the parasite deep inside the body of a mouse, researchers used a modified camera that had been originally used in astrophysics to amplify light from very faint stars.

In order to control such pests as mosquitoes, it is important to understand how they reproduce. In most species of insects, birds, and some arachnids, the second male to copulate with a female fathers most of her offspring. How does the sperm of the last male to mate maximize its chances of being fertilized? Researchers from the University of Chicago provided some answers by studying fruit flies. The female fruit fly mates with multiple males, storing the sperm in three specialized storage organs where it remains until it is required to fertilize her eggs. However, the chances of becoming a father are not equal for all the males; the last one to mate with the female tends to father the most fruit flies. By labeling the sperm of the first mate with GFP, researchers were able to distinguish between sperm released by the first and second partners.[23] (See figure 7.) They found that the first male's sperm is displaced by that of the second partner. However, they could not find what happened to the sperm of the first male.

Besides being displaced by the sperm of the second male, the sperm of the first mate is also incapacitated by that of the second partner. This effect is more pronounced the longer the sperm is stored in the seminal receptacle of the female. At this point, no one knows why the sperm from the second mate nearly always beats out that from the first one. Is there an evolutionary reason? This has yet to be determined.

GFP is used in thousands of experiments every day, a large number of them don't work and most never make it into the scientific literature. Very few are ever reported in the popular press; only the ones that make good copy are. They are rarely the most important ones. Nonetheless, they illustrate that GFP is becoming a commonplace scientific tool—just like a microscope, an electrophoresis gel, or even a test tube.

CHAPTER 8

ALBA, THE FLUORESCENT RABBIT

One of the many indicators that the green fluorescent protein revolution had begun was when GFP made the transition from science to art. It became the paintbrush of a new form of art—transgenic art. According to Eduardo Kac (pronounced *cats*), an artist and professor at the Art Institute of Chicago, the goal of transgenic art is to raise awareness of genetic engineering. He suggests, tongue-in-cheek, I hope, that transgenic artists can help increase biodiversity by inventing new life-forms.

Normally it is impossible to see whether organisms have been genetically modified, but when they have been transgenically modified with GFP, the modifications are made fluorescently obvious. To date there are not many artists using GFP in their works, and Eduardo Kac is by far the most famous of them all. A search through the popular media for references to GFP reveals that more than half the articles are about Kac's art. The myriad uses described in the scientific literature are rarely reported in the mainstream press.

Kac was born in 1962 in Rio de Janeiro. During the eighties, he protested the Brazilian dictatorship by giving "performance art" demonstrations on Ipanema Beach, reciting porn poems while wearing a pink miniskirt. In 1989 he moved to Chicago, where he received an MFA from the Art Institute of Chicago. He never studied biochemistry, molecular biology, or chemistry and uses biotech labs to produce his transgenic organisms. He is currently the chair of the art and technology department at the School of the Art Institute of Chicago.

Kac has produced two exhibits based on GFP technology, "GFP Bunny" and "The Eighth Day." They are both part of his "Creation Trilogy." The first part of the trilogy is "Genesis," which doesn't involve GFP but will be described due to its relationship with the other two pieces.

In "Genesis" Kac translated Genesis 1:26, "Let man have dominion over the fish of the sea and over the fowl of the air and over everything living that lives upon the Earth," into Morse code. Since both DNA and the Morse code are made up of four different characters, Kac was able to convert the dots, dashes, and spaces between letters and words in the Morse-coded version of this passage into the DNA nucleic bases C, T, G, and A, respectively. He then hired a biotech company to synthesize the "Genesis gene," which was injected into fluorescent bacteria. The gene is an artificial one and probably not expressed in the bacteria. Visitors to the exhibit saw a projected view of the bacteria if and only if they switched on a UV lamp that briefly irradiated the bacteria and mutated them, thereby rewriting Genesis 1:26. The exhibit was shown at galleries in Linz, Sao Paulo, Chicago, New York, Yokohama, Athens, Madrid, and Pittsburgh. For each show, a new "Genesis gene" was created. In some cases, a Web link to the exhibit was established, and Web surfers were given the opportunity to view and thereby mutate the bacteria. The Genesis exhibit premiered at Ars Electronica in 1999, and after the show, Kac took the mutated bacteria back to the lab and had the modified "Genesis gene" sequenced, converted to Morse code, and translated back to English. This was not the aim of the exhibit and was not done after most shows, which lasted between a few weeks and a couple of months. Most of the mutations were nonsense mutations, but some made sense and were interesting; for example, *fowl* was mutated to *foul*. In Genesis, Kac has tried to break the barriers between art and life.

Alba is a cuddly albino rabbit that hops around, snuffles its nose, and munches carrots just like any other rabbit. Turn off the lights, switch on the ultraviolet lamps, and it becomes GFP Bunny, a transgenic artwork. Alba, Spanish for "dawn," is both alien and cuddly. Turn on the UV light and she changes from loveable family pet to a disconcerting vision of the future, a science fiction pet with an eerie green glow emanating from every cell—from her paws, her whiskers, and especially her eyes. (See figure 8 in the photo insert.)

Alba was created by Louis-Marie Houdebine of the French National Institute for Argonomic Research. There is some disagreement about the origin of the initial idea to create Alba. Producing a transgenic rabbit with GFP was his own idea, Kac claims, and Alba was created for his exhibition. Houdebine disagrees—he says that he had produced numerous transgenic GFP rabbits before being approached by Kac.[1] In fact, in 1998 his team took some commercial GFP and injected it into the eggs of three albino rabbits. When the rabbits reached maturity, the researchers bred them and raised the offspring that showed signs of incorporating GFP; the non-GFP offspring were destroyed. Over 150 transgenic GFP bunnies were bred in this way, which was more than the hutches could handle, so they were culled back. Now the permanent population of transgenic GFP rabbits at the French National Institute for Argonomic Research is fewer than fifteen. According to Houdebine, Kac came to the research center and picked out a bunny for his show.[2]

Alba, the GFP rabbit from the French National Institute for Argonomic Research, is part of the second work in the "Creation Trilogy," but there was supposed to be more to the GFP-bunny piece than just Alba—the dialogue created by the pet/alien dichotomy and the social integration of Alba were important parts of the exhibit. Alba's public debut was scheduled for an exhibition of digital art in Avignon, France. Kac and Alba were going to live in a faux living room created in the gallery, signifying how biotechnologies are entering our lives, even in the privacy of our living rooms. However, on the eve of the show, the director of the institute who had created Alba reportedly refused to release her to Kac. This fueled the dialogue portion of the exhibit, and soon Alba was competing with the Olympics for headlines in the *Boston Globe*, *Le Monde*, the BBC, and ABC news. The GFP-bunny exhibit was meant to be a political project that would break down the barriers between art, science, and politics, and in this it succeeded. For many people, their fears of genetically modified organisms, the human genome project, and cloning were realized when they saw the strangely fluorescent eyes of Alba. The GFP bunny demonstrated how unregulated the biotechnology sector is and how easy it is to produce mutant species or transgenic organisms whose existence many of us would find disturbing and perhaps even unethical. One has to remember that Kac did not "make" Alba; she is a product of science. He used her as a symbol for all transgenically modified organisms and of what is possible with biotechnology; she was meant to be provocative. Beside having many detractors, Kac's project also has many supporters. Robert Silverberg, the science fiction writer, has entered the GFP-bunny debate. He would like to know why scientists can create transgenic organisms while artists can't and whether breeding a "phosphorescent [sic] rabbit" is any sillier than breeding a dachshund.[3] Brigitte Boisselier of the Raelians, a group that has made headlines by claiming to be the first to clone human babies, has also got in on the act by proclaiming, "This glowing rabbit is just great. . . . An artist thought about it, a scientist did it . . . and the whole community is complaining."[4]

Two years after Alba first made the headlines, she died at four years of age—or maybe she didn't die. According to her creator, Louis-Marie Houdebine, he was informed that Alba, or GFP.014 as she was known to him, died in July 2002. She was four years old, which is a normal life span for rabbits at the French National Institute for Argonomic Research. Kac reportedly is not convinced that Alba is dead; he apparently believes that Houdebine has falsely declared Alba dead in order to put an end to all the unwelcome publicity he is getting. He says that a typical rabbit lives up to twelve years and that Alba was specifically made for him in January 2000, which made her two years old and not four.[5] Kac has not given up on getting his fluorescent bunny to Chicago and still has a Web site to "free" Alba.*

Kac modified some amoeba, fish, mice, and plants by adding the GFP gene to their genome and placed them in a clear four-foot-diameter plexiglass dome. The exhibit, a transgenic biosphere called "The Eighth Day," is the last part of the "Creation Trilogy." While most transgenic organisms have been developed in isolation, "The Eighth Day" dome is meant to

---

*http://sprocket.telab.artic.edu/ekac/bunnyadd.html.

symbolize a new ecology that is forming between genetically modified crops in the United States. It is the eighth day in the creation of man and earth. The centerpiece of the display is a robot that is driven by green fluorescent amoebae called *Dyctiostelium discoideum*. When they are active, the robot goes up; when they are quiet, it goes down. The robot also has a camera attached to it that can be controlled by Web participants.

Many people may ask whether Kac's GFP displays are art or whether it is ethical to develop transgenic organisms solely for their shock or artistic value. In fact, I suspect that is what the artist wants us to do. He wants us to think about genetically modified organisms and where the borders should be drawn. I think there is no doubt that he has succeeded in instigating such a debate about genetically modified organisms.

In October 2002, the Exploratorium, which is located inside the Palace of Fine Arts in San Francisco's Marina District, displayed some transgenic *C. elegans*. They were part of the "Traits of Life" exhibit, which showed works of cutting-edge research. Martin Chalfie produced the roundworms with GFP in their nerve cells. There was no controversy about the exhibit. Was it because this was considered science, not art, or because a tiny worm doesn't produce the same emotional response as a fluffy rabbit?

Just flip through the pictures in the photo insert of this book and you should not be surprised that other scientists with artistic ambitions and abilities are converting their research into art. Not the transgenic art of Eduardo Kac, but the more conventional photographic and cinematographic art we are more used to seeing.

Transgenic Light was a collaborative experiment among some Stanford researchers that explored the aesthetics of images produced using green fluorescent protein. According to the artists, "Transgenic Light sought to map the rapidly advancing zone of mediation opened up by GFP—a realm that connects several converging strata, including the discursive and the phenomenal, the semiotic and the material, the digital and the organismal, the natural and the artificial, and the scientific and the aesthetic."[6] The installation was on display from June 12 to August 25, 2002, at Stanford University's Cantor Center for the Visual Arts, and the exhibit had three displays. Professor Bruce Baker of Stanford University had donated a stereo fluorescent microscope with a camera so that viewers could observe a live video feed of fruit flies whose eyes had been genetically altered to express GFP on one screen. At the same time, a montage of GFP images from recent science experiments was being displayed on another large screen. The final component of the show was a five-and-a-half-minute digital movie that presented "a reinterpretation of GFP image data as embodied natural landscapes."[7]

At the last two meetings of the European Life Science Organization, a competition for the best multimedia presentation has been held. In the *Cinema of the Cell* session, it was not the quality of science that was being evaluated; it was the artistic merit of the presentations that was judged. More than a dozen young cell biologists showed their best film clips.*

Artistic films of this type are nothing new. Movies of microscopic phenomena have been around for more than one hundred years. In 2000, the Museum of Modern Art in New York City featured a thirteen-minute film *Le Mouvement des plantes* as part of a festival celebrating French films. The film was made by Jean Comandon and Pierre de Fonbrune, two pioneers of microcinematography, who both died long before fluorescent proteins were used to light up the microscopic and submicroscopic research world.[8]

Most of the entrants in the 2002 and 2003 *Cinema of the Cell* sessions used glowing genes in their films. Remi Dumollard, a postdoc at University College London, has been using fluorescent proteins to visualize calcium oscillations. In the film, which he entered in the competition, he used all kinds of filters to make the colors more intense and added music to it, so that it had the same rhythm as the calcium waves. According to Holger Breithaupt, news editor at the European Molecular Biology Organization (EMBO), the result is striking because the viewer is left with the impression that the cell is pulsing with the rhythm and not vice versa.[9]

The GFP revolution in art is upon us.

*Some of these clips can be viewed at http://www.bioclips.com.

CHAPTER 9

LIGHT IN A CAN

What do GFP and luciferase look like? We know that GFP is green fluorescent and that luciferase gives off a flash of light in the presence of luciferin, ATP, and oxygen, but what does the actual protein look like? Proteins have complex three-dimensional shapes that are determined by their amino acid sequences. Because GFP and luciferase always have the same sequence of amino acids, no matter what type of organism they are produced in, they will always have the same shape. This is significant because it is the three-dimensional structure of proteins that is responsible for their characteristic properties. So in order to understand GFP's behavior, it is very important to know its structure; the same is true for luciferase. Proteins are so small that there is no way they can be seen under a microscope. Earlier we saw that Doug Prasher determined the amino acid sequence of the 238 amino acids that make up GFP. Ten years ago, scientists were hoping that determining the human genome and the genomes of other species would answer many of the big questions in science and move medical research ahead in leaps and bounds. We now have most of the information encoded in the human genome—we know the human recipe book. However, that hasn't helped us as much as we had

113

hoped it would. The genes tell us about the sequence of amino acids in the proteins they code for, but they tell us nothing about how the proteins interact with each other. This is the field of *proteomics*, a new and very important area of research, which often requires a knowledge of the shape that the proteins adopt. Many research groups are trying to use computers to calculate three-dimensional protein structures from their amino acid sequence. Unfortunately, this is not possible yet. The process of protein folding is still a mystery that is waiting to be solved. It is an important one, with many consequences in science and medicine. If the proteins misfold, they no longer function properly, and diseases like mad cow's and Alzheimer's can result. Protein folding normally takes about one thousandth of a second; this is extremely fast, but the process is so complicated that even the fastest desktop computer takes at least a day to simulate one billionth of a second of the folding process. This means it would take one computer about a million days, or three thousand years, to model the folding of a protein. There are hundreds of computational chemistry groups working on the protein-folding problem, trying to solve it by simplifying the problem and using major computational power. It is a challenge that ranks as one of the toughest problems in biology. Two interesting and very different approaches are being used to tackle the problem of protein folding.

On December 6, 1999, IBM announced a new five-year $100 million research initiative to build a new supercomputer that was five hundred times as fast as the fastest computers existing at the time. The computer, called Blue Gene, would be used to work on trying to predict how proteins fold. It would be one thousand times as powerful as Deep Blue, the computer that beat world chess champion Gary Kasparov in 1997, and about two million times faster than a 1999 desktop PC. According to an IBM press release, Dr. Paul M. Horn, senior vice president of IBM research, justified the expenditure to IBM stockholders in the following way:

> This is exactly what IBM Research does best—continuously placing big, aggressive bets on technologies that change the future of computing. In many ways, Deep Blue got a better job today—if this computer unlocks the mystery of how proteins fold, it will be an important milestone in the future of medicine and healthcare. Breakthroughs in computers and information technology are now creating new frontiers in biology. One

day, you're going to be able to walk into a doctor's office and have a computer analyze a tissue sample, identify the pathogen that ails you, and then instantly prescribe a treatment best suited to your specific illness and individual genetic makeup.[1]

We are still far many years from that point. To date, the best the IBM supercomputers can do is figure out how a string of twenty amino acids called the *trp-cage* will fold.[2] It's still a long way from being able to predict how a small protein like GFP, with 238 amino acids, will fold and even further from predicting how a larger protein, like the 550–amino acid luciferase, will adopt its final active structure; however, it is an important start.

Folding@home is not a laundry company that specializes in folding clothes. Rather, it is an attempt to make use of the computational capacity that is being wasted when home personal computers are sitting around not being used. Anyone can download folding@home software from the Web. It is free and looks just like a screensaver to the user. As soon as the computer has been sitting idle for a predetermined time, the folding@home software kicks in and starts calculating. Any keystrokes stop the calculation and the molecule disappears from the screen. Since October 2000, more than four hundred thousand people have downloaded the folding@home software, and it now runs on an average of almost one hundred thousand computers around the world every day. Buying and running one hundred thousand computers would cost about $50 million and would be a nightmare to look after. According to Professor Vijay Pande, who is in charge of the project and is based at Stanford University, the program is downloaded mainly by people with interests in computers, biology, and fighting diseases, as well as teachers who find that folding@home is a unique way to get students interested in science. "This is like having a whole new kind of funding agency for research—namely, the general public donating its computers," said Pande. "When you factor in the maintaince they are doing, the operating system upgrades, and so on, that's a gigantic resource!"[3] Using the power of distributed computing the folding@home group has managed correctly to predict the folding of a small twenty-three–amino acid sequence called BBA5.[4]

Even with all the computational resources available to IBM and the four hundred thousand computers used in the folding@home project, we are still far away from simulating the folding of real proteins such as aequorin, luciferase, and green fluorescent protein. So how can we determine the three-dimensional structure of these proteins? At this point, the only way to determine the structure of a molecule that is the size of GFP is by x-ray crystallography.

The first x-ray structure was solved in 1913 by a father-and-son team, William Henry and William Lawrence Bragg. They solved the structure of sodium chloride and in 1915 received the Nobel Prize in Physics. Lawrence was twenty-five at the time, the youngest-ever Laureate. Many people assumed that most of the work had been done by Henry Bragg, but it was actually Lawrence Bragg, the son, who came up with all the theories. And Henry tried unsuccessfully to ensure that his son got the appropriate recognition. A few days before his death in 1971, Lawrence wrote to his friend and Cambridge University crystallographer Max Perutz, "I hope that there are many things your son is tremendously good at which you can't do at all, because that is the best foundation for a father-son relationship."[5]

X-rays have wavelengths that are the same order of magnitude as the distance between bonded atoms. Because of this, complex diffraction patterns result when crystals are irradiated with x-rays. The position and intensity of the x-rays that have bounced off the atoms depend on the size of the atoms and their position relative to all the other atoms. Diffraction patterns can be calculated based on hypothetical molecules. When the calculated and observed structures are identical, the crystal structure has been solved. The process is iterative and very tedious. Before the advent of computers, it took months of calculations to solve the crystal structure of small (about a hundred atoms) structures; now it can be done in one afternoon. Proteins are extremely large molecules; without computers it would have been impossible to solve the crystal structure of proteins. The first one solved was the structure of myoglobin, which was determined in 1960 by Max Perutz and John Kendrew. Perutz started studying hemoglobin with Sir Lawrence Bragg in 1937, and ten years later, he was joined by Kendrew, who looked at crystals of the related muscle pigment myoglobin. In 1962 Perutz and Kendrew were award the Nobel Prize in Chemistry for their crystallographic work. That same year, the Nobel

Prize in Medicine went to Francis Crick, James Watson, and Maurice Wilkins for discovering the double helical structure of DNA. Their ideas were largely based on x-ray diffraction patterns of DNA obtained by Rosalind Franklin.

Often the biggest problem in protein crystallography is growing crystals that diffract x-rays well. Osamu Shimomura and others grew beautiful crystals of GFP many years ago; unfortunately, these crystals did not diffract well, and the GFP community had to wait a long time before the GFP structure was determined. In 1996 two independent groups published the three-dimensional structure of GFP at the same time. George Phillips's group from Rice University solved the structure of the wild-type GFP, and Steve Remington and coworkers published the enhanced GFP structure.[6] Wild-type GFP has the same amino acids sequence as GFP found in the jellyfish; enhanced GFP is a mutant of GFP in which the 65th amino acid has been changed to make a mutant of GFP that was more fluorescent than the wild-type GFP. Enhanced GFP was discovered by randomly making very small changes in the gene of GFP and expressing the proteins encoded by the mutated genes. If any of the GFP mutants had desirable properties, they were isolated and their sequence was determined. This technique is called *random mutagenesis*, and it is very effective at generating new and useful GFP mutants because it is very easy to pick out the desirable GFP mutants that fluoresce brighter, quicker, or with different colors. In 1996 when the two crystal structures were solved, enhanced GFP, often abbreviated EGFP, was more commonly used than the wild-type GFP.

Both the GFP and enhanced GFP structures were essentially the same. Figure 9 in the photo insert shows the crystal structure of GFP. It has a unique barrel-like structure, with eleven staves surrounding the three amino acids that compose the chromophore, which is responsible for GFP's fluorescence. The GFP barrel has a diameter of about 24×10^{-10} meters and a height of 42×10^{-10} meters. Phillips has coined the phrase "light in a can," which very aptly describes the structure of GFP.

A new mutant was constructed based on information obtained from the GFP crystal structure. The 203rd amino acid was changed from a threonine to a tyrosine because it looked as if the tyrosine would stack right above the chromophore, thereby changing its fluorescent properties.

Indeed the resultant mutation visibly changed the GFP—it no longer fluoresced green; instead, it gave off yellow light. The mutant is called *yellow fluorescent protein* (YFP).[7] Without solving the crystal structures, it would have been impossible to predict that the 203rd amino acid would be lying right above the chromophore, which is made up of amino acids 65, 66, and 67, and there probably wouldn't be a yellow fluorescent mutant.

Blue, cyan, and citrine GFP mutants have also been produced and are commercially available. There is even a class of "gold" GFPs that has been designed by exchanging some of the amino acids in the chromophore with an amino acid that does not occur naturally.[8]

There are other uses for knowing the three-dimensional structure of GFP besides designing new mutants of GFP. For example, this knowledge has been used in conjunction with some other experimental data to propose a more detailed mechanism for the formation of the chromophore. Once the 238 amino acids of GFP are made in the cell, they fairly rapidly fold into the barrel shape observed in the crystal structure. Amino acids 65, 66, and 67 are arranged in a very tight turn right in the center of the can, which holds them firmly in place, not allowing them to move much. Then, slowly the amino acid 67 attacks 65 to form the chromophore and make GFP fluorescent. One can unfold proteins from their characteristic three-dimensional shapes by heating them or by adding certain chemicals. This is called *denaturing the protein* and usually results in the protein losing its ability to function. If GFP is unfolded (denatured), it absorbs light just like the folded form but does not fluoresce. That is because the chromophore is still present in the denatured form, but it needs the amino acids around it to make up the barrel to fluoresce. Synthetic GFP, all 238 amino acids, has been made in the laboratory—it behaves identically to genetically expressed GFP. It folds into the barrel shape; the chromophore is formed by amino acid 67 attacking 65, and it fluoresces if irradiated with ultraviolet light. Two amino acids can be removed from the beginning and six from the end of the GFP sequence without losing fluorescence. If any more are removed, the barrel is no longer formed, the chromophore is not formed, and no fluorescence is observed.

Knowing the three-dimensional structure of GFP has allowed chemists to do molecular origami with GFP. An extraneous loop made up of twenty amino acid residues was inserted between residues 157 and 158 of GFP. These residues are located in a loop on the surface of GFP. The resultant mutant was fluorescent, and one can therefore assume that the new modified GFP with its 258 amino acids was still able to fold into the characteristic barrel shape. But there was some stuff (twenty amino acids) hanging off the side. Cyclic GFP has been made by using ten amino acids to link the beginning and the end of GFP to each other. Normally, GFP is a string of 238 amino acids folded up into a barrel shape. Cyclic GFP is an elastic band of amino acids folded into a barrel shape. GFP is so useful because it is stable—cyclic GFP is even more stable than GFP, and it is difficult to denature cyclic GFP. One of my favorite examples of molecular origami is an experiment in which the gene for GFP was cut in two so that two halves of GFP were expressed in the same cell. The two halves did not join up and form a fluorescent GFP barrel. However, when a series of leucine amino acids were added to the end of the first GFP half and the beginning of the second GFP half, a fluorescent GFP barrel was formed. This is because the leucines act like a zipper, joining the two halves so that they can fold correctly and form the GFP barrel and the chromophore.[9] There are many cells and parts of cells that scientists would like to label; however, this is not always possible because often they do not know of any genes that are exclusively expressed in the cells and organelles of interest. In 2004 Marty Chalfie, whom we met in chapter 7, reported that his group had developed a system that used the reconstitution of two GFP fragments with an antiparallel leucine zipper to label cells and organelles. By choosing two sets of promoters whose expression patterns overlap for a single cell type or organelle, the Chalfie group was able to produce animals with fluorescence only in those cells. This was done by modifying a promoter so that it started the production of half a GFP with part of a leucine zipper, and another promoter was created to express the other half of GFP with its complement of the leucine zipper. In order for fluorescence to be observed, both halves of GFP had to be expressed in the same local; in this way, fluorescent GFP was restricted to cell-types and organelles, which only contained both promoters.[10]

In October 2004, the Nobel Prize in Chemistry was awarded Aaron Ciechanover, Avram Hershko, and Irwin Rose for their finding that unwanted or problematic proteins in living cells are tagged with a chain of ubiquitin molecules, thereby marking the proteins for degradation by organelles called *proteasomes*. The ubiquitin molecules are the kiss of death to the protein—once an undesired protein has been tagged with ubiquitin molecules, it cannot escape being sliced and diced by the proteasome. It is very difficult to examine ubiquitination in living cells, in part because only a small subpopulation of each protein is ubiquitinated at any one time. In the same month as the 2004 Nobel chemistry prizes were announced, Tom Kerppola at the University of Michigan Medical School developed a GFP-based method for observing proteins as they get the ubiquitin kiss of death; this method is also based on the leucine zipper. He tagged ubiquitin with half a GFP molecule and the proteins on "death row" with the other half of the GFP. When the labeled proteins were ubiquitinized, the two halves of GFP were joined and fluorescence could be observed.[11]

Since all GFP mutants have the same barrel shape as that of wild-type GFP, it has become much easier to determine the crystal structures of the mutants based on the structures of GFP and EGFP. Forty-eight crystal structures of GFP and GFP mutants have been published. Like all other three-dimensional structures of molecules with more than five hundred atoms that have been published in the scientific literature, they have been deposited in an electronic database known as the protein data bank.* On March 9, 2004, it contained 24,615 structures, most of them determined by x-ray crystallography. Two of the GFP structures deposited in the protein data bank do not contain the chromophore and do not fluoresce. These two structures are extremely interesting to me because in 1996, before the first crystal structures of GFP were published, we predicted that an amino acid called arginine had to be very close to the 65th amino acid before it could attack the 67th amino acid to form the chromophore. When the crystal structures of GFP were reported in the literature, we were delighted to see that the 96th amino acid, which is an arginine, was very close to the 65th amino acid, just as we had predicted. The two structures I mentioned earlier are structures in which the arginine at position 96 has been replaced with an alanine—in both structures, the 65th amino acid did not attack the 67th one. There is no chromophore, and the GFP mutant does not fluoresce. This is important because it provides us with information about the unique structural properties that allow GFP to

---

*http://www.rcsb.org/pdb/.

attack itself to form the chromphore, which is responsible for its fluorescent properties.

The first crystal structure of firefly luciferase was solved in 1995 and published in 1996.[12] The 550 amino acids that make up luciferase do not form an easily recognizable shape like that found for GFP. Instead, they form two separate blobs—a large blob made up of the first 436 amino acid residues, and the second comprises the last 104 residues. The crystal structure includes neither the ATP nor the luciferin molecules. However, based on other experiments, researchers have worked out that the molecules bind to the larger blob of luciferase and that the smaller blob rotates once the luciferin and ATP have bound, effectively shutting them inside the luciferase protein. Based on the luciferase crystal structures, molecular biologists have been able to produce luciferase mutants that give off light in a variety of colors.

We will next see some of the practical consequences of knowing the three-dimensional structure of GFP.

CHAPTER 10

RED SHEEP FROM RUSSIA

After Chalfie's first reported use of GFP, it took less than three years for scientists and biotech companies to realize the enormous potential that GFP held. In order to maximize its usefulness, scientists designed many mutants. One of the most popular ones is enhanced GFP (EGFP), which is a mutant that has a more intense fluorescence than wild-type GFP. *Aequorea victoria*, the jellyfish from which GFP was isolated, is found in the cold Pacific Ocean, and as a consequence, GFP most efficiently folds into its barrel shape at 15°C (59°F). Since most organisms studied with GFP have much higher body temperatures, a new mutant was devised that optimally folds at much higher temperatures; it is called *cycle 3*. Yellow, blue, citrine, cyan, and gold GFPs have been formed by selective mutations. If one changes the 66th amino acid in GFP, which is normally a tyrosine, to an amino acid called histidine, the green fluorescent protein no longer gives off green light; instead, it emits blue light, and the mutant is called *blue fluorescent protein*. To date, despite much effort, no one has been able to mutate GFP to give off red fluorescence. Many researchers have been trying to find red mutants because most animal tissue is nearly transparent to red and infrared radi-

123

ation, so if a red mutant were made, it would be more readily detected in vivo. In a dark room, cover a flashlight with your hand, and you will see a reddish glow emitted through your hand. This is a common example of the ease with which red light is transmitted through animal tissue.

At the same time as these mutational studies were taking place, a huge effort went into finding other proteins that had GFP-like properties. Bioluminescence is a very common phenomenon in the oceans, particularly in the deep seas. Surely some of the bioluminescent sea creatures would use an aequorin-GFP system in the same way that *Aequorea victoria* does. The sea pen, *Renilla reniformis*, emits a green fluorescence very similar to that observed in *Aequorea*. However, the protein responsible for the green fluorescence has not proven to be very useful. And none of the mutants of *Renilla*-GFP have been found with red fluorescence. Prior to 1999, many other organisms were studied that had similar mechanisms for producing bioluminescence to the one found in *Aequorea*. None of them had any properties, such as red fluorescence, that made them as useful as GFP is as a marker molecule.

At the time, the consensus of opinion was that there weren't many GFP analogs out there among all the other species. We now know that that was completely wrong. The reason no one could find GFP-like proteins in other species is that they were all looking at the wrong species. *Aequorea victoria* is not brightly colored or fluorescent; in fact, it so colorless and transparent that it is commonly called the crystal jellyfish. The jellyfish gives off a flash of green bioluminescence only when mechanically stimulated. Therefore, researchers who were looking for GFP analogs focused on species that were similar to *Aequorea victoria*, species that emitted blue or green bioluminescence. Coloration was considered unimportant. This strategy wasn't a total failure, as some GFP-like proteins were found, but the actual breakthrough did not occur until the paradigm of the search was changed.[1]

Mikhail Matz and Sergey Lukyanov, researchers at the Russian Academy of Sciences, made the breakthrough in finding GFP analogs by choosing a different path from that of all their competitors. Since 1994, when GFP was first used as a tracer molecule, analogs to GFP were being sought among bioluminescent systems that had two proteins involved in producing fluorescence, the idea being that GFP analogs would have evolved in the same way as GFP, which takes the short wavelength light from aequorin and converts it to longer wavelength green light, which it then emits. Lukyanov and Matz were intending to screen all the same species that most scientists were studying, until they spoke about their project to Yulii Labas of the Institute of Ecology and Evolution of the Russian Academy of Sciences. He was an expert in the evolution of bioluminescent systems, and he pointed out that many of the species that Matz and Lukyanov were planning on examining had recruited their bioluminescent proteins from various biochemical pathways that had nothing to do with light production. Maybe GFP was also a "recent recruit," and perhaps some of its analogs could be found in species that fluoresce but do not bioluminesce. He continued, "For instance a friend of mine has a reef aquarium with many representatives of Anthozoa class, and although none of these species possess bioluminescent capabilities, they are very green-fluorescent! Why not free your minds and try these?"[2] So Matz and Lukyanov decided to take their hunting expedition for GFP analogs a little further away from jellyfish. On Labas's advice, they reasoned that GFP-like proteins were not necessarily linked to bioluminescence. The physics of bioluminescence is the same throughout the animal kingdom, though the actual chemistry differs significantly. This made them think that bioluminescence evolved relatively recently by combining and modifying preexisting proteins. If this idea were correct, then perhaps some of GFP's evolutionary ancestors could still be present in nonbioluminescent organisms. They would most likely be brightly colored and naturally fluorescent, but not bioluminescent.

In case you are confused at this point—I know that I was when I first encountered "nonbioluminescent, brightly colored, and naturally fluorescent" all in one phrase—here is an explanation. The angler fish discussed in the first chapter is bioluminescent; it is giving off its light so that we can see it; fireflies and jellyfish are also bioluminescent, as they can give off luminescence (light) whenever they have the desire. The large star colony shown in figure 10 in the photo insert is not bioluminescent; it does not give off any luminescence. However, it is fluorescent, as it does give off bright yellow light after it has been irradiated with ultraviolet light. *Aequorea* is bioluminescent because it can give off green light by adding calcium to aequorin, which then passes on energy to GFP, which, in turn, gives off the green luminescence. GFP by itself is not bioluminescent, but it is fluorescent. An everyday example of fluorescence is laundry detergent that uses fluorescence to make clothing look whiter by absorbing ultraviolet light and then fluorescing in the visible spectrum.

Lukyanov and Matz decided to start their search for GFP-like proteins in the coral reefs that are native to the Indo-Pacific Ocean. The wide variety of colors and fluorescence found in coral reefs made them very attractive candidates, especially since the pigments responsible for the fluorescence in corals were unknown at the time; perhaps one of them was a GFP analog. Many corals have bright fluorescent colors. The most common color among corals found in shallow reefs is green, presumably because it can provide protection from the intense tropical sunlight. For corals found in deeper waters, blue is more common because the low-energy sunlight has been filtered out by the seawater, leaving blue light. Even though most corals are blue or green, many other colors could be found.

Having decided to look for GFP-like proteins in corals, they had to develop a method for finding out whether there were any GFP analogs present in the corals. Using the isolation and purification methods that were applied to find GFP in the jellyfish was not a very attractive option. It would require lots of painstaking and tedious work to isolate and then characterize the protein responsible for fluorescence in the corals, and there was no guarantee that the protein would be a GFP-like protein. Even if it were a GFP analog, it might not possess any properties that would make it a good tracer molecule. So instead of going with the older painstaking methods that Shimomura had used to find GFP, Lukyanov used his intuition and modern biochemical methods. He looked at the crystal structure of GFP and decided that the most crucial amino acids in GFP were the ones that were modified to form the chromophore and the ones that were located at the tight turns between the staves of the GFP barrel. If there were any GFP-like proteins in corals, they would have to be about the same size as GFP and have the same amino acid sequence as GFP in these crucial areas. Lukyanov then used the same techniques that Prasher used to find the GFP gene in *Aequorea*. He chose five brightly colored coral species from the Indo-Pacific area, multiplied all the DNA from the colored body parts, and went searching through them to find any proteins that had the correct amino acid sequences for the tight turns and the chromophore. Lukyanov and his coworkers found six proteins that met his requirements. They were cloned and expressed. Lo and behold, they were fluorescent and had similar photophysical properties to GFP. One was red, another was yellow, and the remainder were green. The Russian researchers were also able to express the GFP-like proteins in mammalian and frog cells, showing that they could be used in a similar fashion to GFP.[3]

If we go back to the cookbook analogy, Lukyanov's method was the same as deciding that any GFP-like cakes need four eggs, cinnamon, and three hours of baking at 320°C, and then going through all the cookbooks and looking for cakes that have four eggs, cinnamon, and three hours of baking at 320°C. Once the recipes were found, they were used for making the cakes to see if they tasted like a GFP cake.

Although the coral fluorescent proteins have only 26 to 30 percent sequence identity with the amino acid sequence of *Aequorean* GFP, they have the same barrel structure as GFP and a very similar chromophore. One can therefore presume that they have the same evolutionary origin and that bioluminescence evolved from fluorescent proteins.

William Ward of the Center for Research and Education in Bioluminescence and Biotechnology at Rutgers University, mentioned earlier, has written a very interesting and amusing article titled "Fluorescent Proteins: Who's Got 'Em and Why?" He wrote, "And finally, the bioluminescence community concluded, without a shred of evidence, that GFP is found only in bioluminescent coelenterates. . . . We were unaware that marine and physical scientists were concurrently studying the fluorescence of corals. They were just as unaware of us."[4] It took a research group based in Moscow, thousands of miles from the habitat of either *Aequorea* or corals, to make the connection and to search for GFP-like proteins in coral reefs.

The red GFP-like protein from the corals has proven to be very useful and is commercially available. It is known as DsRed because it is found in *Discosoma striata*. The crystal structure of DsRed has been solved and its properties have been thoroughly studied. The red fluorescence that is observed in mature DsRed does not appear immediately; instead, DsRed first gives off weak green fluorescence before slowly converting to a red form. Perhaps GFP is a broken form of DsRed that was developmentally arrested at the green fluorescent stage. By changing amino acid 83, DsRed's development can be stopped at the green fluorescent stage.

A year after the discovery of DsRed was reported in the literature, two groups simultaneously solved the crystal structure of DsRed.[5] Since Lukyanov and Matz used structural elements of GFP to find GFP analogs, it shouldn't be surprising that DsRed has a very similar barrel-shaped structure to that observed in GFP. The barrel has about the same dimensions as that found in GFP, both have eleven staves, and both enclose a centrally located chromophore. One big difference between the structure of GFP and DsRed is that DsRed is a *tetramer*, which means that it does not like to be by itself. While GFP barrels are normally not associated with other GFP molecules, a DsRed will have three other DsRed molecules associated with it, both when it is in a solid crystal and when it is in a solution. This is a problem, since four DsReds together are a lot larger and more cumbersome than one GFP molecule. DsRed-tagged proteins are more likely to change their behavior due to the larger DsRed aggregates than GFP-tagged proteins. Fortunately, in 2002, Roger Tsien and his group at the University of San Diego managed to make a mutant of DsRed that does not form tetramers.[6] So now we have a glowing gene that codes for a protein that is very similar to GFP in all respects, but it emits red light, which is just what everyone was looking for. Furthermore, DsRed is found in corals that are located in warm tropical waters, while GFP has its home in jellyfish that live in the cold Pacific waters off the coast of Washington and Oregon states. As a consequence, GFP optimally forms at temperatures that are much lower than those used in the laboratory. DsRed is optimally formed at room temperature.

To date, twenty-seven GFP-like proteins have been cloned and characterized in an attempt to find another protein that extends the GFP color palette, which includes green, yellow, orange-red, purple-blue, or nonfluorescent. All have the GFP barrel structure and similar chromophore structures.

Not long after Matz and Lukyanov found GFP-like proteins in corals, it was demonstrated that these proteins were responsible for the color of reef-building corals. This is a bonanza for evolutionary scientists because it means that the color of the coral is essentially determined by the sequence of a single gene. In most other colored organisms, the color is due to a pigment, which is made by a series of enzymes, and it is therefore very difficult to study the evolution of differently colored pigments. Questions about the evolution of the amazing color diversity found in corals can now be addressed by studying the evolution of the molecular sequences. This is much easier than studying color diversity in flowers, where many genes are responsible for the color. We hope to get answers to questions like, Does the color diversity in corals arise from a diversity of functions that the GFP-like proteins have, or is it solely due to random variations? Labas, Lukyanov, and Matz have got some initial results that suggest at least some of the different colors in corals have arisen due to natural selection and that they have specific functions. At the same time, it seems almost certain that random mutations also have a significant role in generating the diversity of colors.[7]

Many organisms in coral reefs are brightly colored, but very little is known about the chemical basis of the coloration. If GFP-like proteins have been found that do not bioluminesce, is it possible that there could be some GFP-like proteins that do not bioluminesce or fluoresce? Yes, it is. Using the same techniques described, Lukyanov found that there are some bright purple GFP-like proteins in the tentacles of some sea anemones that do not fluoresce or bioluminesce. Through evolution they are related to GFP and can be made to mildly fluoresce by mutating one amino acid. There are some advantages to having naturally colored tracers. The most obvious one is that the label could easily be detected in the field, as there would be no need for ultraviolet radiation. In the paper reporting the discovery of the purple GFP-like protein in sea anemones, Lukyanov writes, "In an industrial setting, transgenic sheep carrying and expressing a particular chromoprotein could simplify and detoxify the process of producing colored clothing by eliminating the need for noxious chemical dyes."[8] Will our grandchildren get to see red sheep grazing contentedly on purple grass meadows?

I don't think it's at all likely that we will be seeing red sheep frolicking in the meadows of Scotland, New Zealand, or any other country for that matter. But that's okay because there are many other far more important uses for DsRed mutants. Some of these uses were found by accident.

Alexey Terskikh and Irving Weissman were trying to find mutants of DsRed that were brighter and developed color faster than the DsRed found in the corals, when they stumbled on a fluorescent timer. Like most "accidental" discoveries that have become important in science, it took someone with knowledge and insight to see the significance and utility of their accidental discovery. Normally, polymerase chain reaction (PCR) is a very accurate method to replicate DNA. That's why it is used in forensic science, where a miniscule amount of DNA can be multiplied without introducing any mistakes in the sequence of DNA bases, so that it can be used in comparisons with other DNA. Error-prone polymerase chain reaction makes multiple copies of DNA but introduces random errors in the replication process. Terskikh and Weissman took DsRed genes and used error-prone PCR to multiply the genes before expressing them. Since the DNA sequence in a gene determines the amino acid sequence in the protein it codes for, making errors in the DNA will result in proteins with a few mutated amino acids. By using error-prone polymerase chain reaction, it is possible to rapidly generate hundreds of DsRed mutants. When Terskikh and his coworkers did the experiment, they found several mutants that were worth examining in more detail because they were brighter than DsRed or because they formed the fluorescent red form faster. One mutant that they had recorded as being bright green in color on the day of the experiment was deeply red fluorescent on the next day. Thinking they had made a mistake, they regrew the mutant DsRed proteins from their purposely messed-up DNA and were surprised to find that the same thing happened again. The mutant was bright green the first day and turned red in the next twenty-four hours; in fact, it passed through a yellow and orange intermediate stage before it turned red.[9] Terskikh and his research group immediately realized that they had stumbled across something very useful, a fluorescent timer.[10] By attaching the gene for the new mutant, called E5, to the promoter of a gene they were interested in, they could observe when the gene was activated but could also differentiate between genes that had been activated at different times. Green fluorescent areas would be indicative of recent activation, yellow-to-orange regions would be observed in areas where continuous promoter activity was occurring, and red fluorescent cells and tissues would be seen where promoter activity had ceased after an extended "on" period.

In order to see if the fluorescent timer really worked the way they had envisioned it, Terskikh and his colleagues tested E5 in the roundworm *Caenorhabditis elegans* and the frog *Xenopus laevis*. In the roundworm, they attached E5 to the promoter for a heat-shock protein that is produced when the worm is exposed to high temperatures. At room temperature, the E5-transgenic worms were colorless. However, when they were subjected to a heat shock regime of an hour at 33°C, they became green fluorescent. Since the worm was subjected to heat for only an hour, E5 was only produced for an hour as the heat-shock protein promoter was activated for only that period. As the E5 aged, it became yellow, then orange, and finally, fifty hours after the heat shock, most of the E5 was red. The color changes could be observed by eye, and the red-to-green fluorescence ratio changed linearly with time.[11]

To see whether E5 could monitor the history of gene expression and record when promoters were activated and switched off in a system (where promoters in different cells were activated at different times), Terskikh examined the formation of the nervous system of the frog. *Otx-2* is a gene that is needed for the normal formation of nerves in the brain of frogs; it is known that the gene is expressed in some areas of the brain before others. The E5 gene was attached to the *Otx-2* gene and injected into frog embryos at the eight-cell stage. The "fluorescent timer" worked just as expected. The areas of the tadpole brain that were known to produce *Otx-2* first turned green before any other areas of the brain; then, as the *Otx-2* promoter was switched off, these areas turned yellow-orange with time, while other areas fluoresced green in response to their promoter activation.

In 2003 Sergey and Konstantin Lukyanov found a protein that naturally behaved like E5, the DsRed mutant.[12] But its fluorescent timer properties are much faster than those of E5. Now we have two fluorescent timers, one, E5, that can be used to differentiate promoter activation and deactivation that occurs slowly, over a day or so, and another that is faster and changes in a matter of hours.

One of DsRed's drawbacks, namely, the fact that it took a long time to form the fully fluorescent form, has suddenly become an advantage because its use as a fluorescent timer would not have been possible had DsRed become red fluorescent the moment it was expressed. Roger Tsien, who has done amazing things with GFP (and whose name appears many times in this book) has said, "These people, you might say, have figured out how to make lemonade out of a lemon."[13]

Perhaps Matz and Lukyanov started the last red revolution in Russia.

CHAPTER 11

ANDi THE GREEN MONKEY AND A YELLOW PIG

"MARTIAN BABY" MYSTERY SOLVED?

Jane Emma Smith was born in a local children's hospital, a healthy girl, seven pounds and two ounces. Her parents, Bruce and Debbie Smith, were ecstatic. After eight years of trying, they finally had the child of their dreams.

It was not an easy delivery. Debbie had been in labor for fourteen hours, and she was exhausted. So after she held and admired baby Jane for an hour, the infant was taken to the nursery, and Bruce and Debbie caught up with their sleep.

While they slept, their lives were about to change. Not even their worst nightmares could have prepared them for the future.

Jane Emma Smith's skin was slowly turning lime green.

A visitor to the nursery saw this and took a series of photographs, which he immediately sold. Within twenty-four hours, poor Jane Emma Smith had lost her identity; thanks to the late-night talk shows, she was now known as the "Martian Baby."

Jane, now five days old, is a perfectly healthy normal baby—except her skin is a faint lime-green color. No other green-colored babies have been recorded in medical history.

Doctors have been unable to find the origin of her unusual pigmentation. The media and their experts have been wildly speculating. For the last three days, the question on everyone's lips has been, "Did you see the pictures of the green baby? How did that happen?"

This morning, the cause of Jane Emma Smith's green skin tone has been established, but many questions still remain unanswered. Professor Nicholas Evans, a prominent molecular biologist, had a suspicion that he knew the cause of Jane's unusual coloring. He called Jane's pediatrician and asked her to look at the baby's body under an ultraviolet light. To the molecular biologist's great disappointment, his theory was correct—Jane's pale lime-green skin gave off a fluorescent glow when irradiated by ultraviolet light.

Somehow, Jane's body was producing a jellyfish protein called green fluorescent protein, which was responsible for the green coloring. Scientists have been using the protein as a marker in many organisms for more than ten years. Its gene has been inserted in plants, fishes, mice, and even human cells, but never in live humans.

Evans is sure that there is no possibility that the GFP gene was accidentally incorporated into Jane's DNA.

Bruce and Debbie Smith have been trying to have a baby for a long time. Debbie's pregnancy was the result of an in vitro fertilization. Evans speculates that the GFP gene was inserted into DNA of the embryo during the fertilization procedure. It had to have happened very early in the embryo's development because many cells all over Jane's body are producing the GFP. Evans could think of no way in which the protein production could be switched off. Nor could he think of any reason why someone had inserted the jellyfish gene in little Jane Smith.

The local police, the FBI, and the CDC are investigating.

This has never happened, and we hope that it never will happen. But could it happen? Could a transgenic GFP human be made using currently available biomolecular techniques? Unfortunately, the answer is yes. It would be much easier to produce a human transgenic GFP baby than it would be to clone a human. However, it would be a very risky procedure, and we hope no one will ever attempt it. Perhaps the fact that human cloning and genetic manipulation of the human germline (cells that link successive generations, e.g., cells that differentiate into sperm and eggs) is currently banned in nineteen countries will help.1

On October 2, 2000, a transgenic GFP rhesus monkey was born. It was the first time a primate was born carrying a foreign gene. It is estimated that we share 93 percent of our DNA with rhesus monkeys, so if a genetically engineered GFP monkey can be produced, it should also be possible to create human GFP babies.

ANDi, named after *i*nserted *DNA* written backwards, was created to help researchers pursue questions that are difficult to study in simpler models for human beings, such as rats and mice. This was an experiment to see whether it were possible to add foreign genes to primates. Gerald Schatten, ANDi's "father," says that ANDi and his future cousins and brothers and sisters will help us bridge that gap between what we know in the mouse and what we are keenly interested in knowing about human development. Transgenic mice will remain the model of choice, but this experiment shows that transgenic monkeys may take the place of mice in special cases. In a response to some criticism, Schatten and coworkers have written:

> As our close relatives, non-human primates might help to eradicate devastating human diseases by accelerating gene- and cell-based cures discovered in rodent models. However, it is the selfsame relationship that raises two types of uncomfortable issues. Could technological discoveries in non-human primates be unethically extended to people? Can we both respect our close relations while also performing experiments on them? Non-human primates deserve extremely high standards of care, and both scientists and the public expect continuing improvements in primate well-being along with increased emphasis on pain-free and non-invasive investigations.[2]

When ANDi was born, Schatten was professor of obstetrics and gynecology and cell and developmental biology at the Oregon Health and Science University in Portland. He was also associated with the Oregon Regional Primate Research Center. Twenty-four researchers and $13 million in research funding helped Schatten with his research to find effective treatments of disabilities and diseases by using transgenic animal models.

The aim of the project was to test whether foreign genes could be inserted into monkey eggs and whether the eggs were still viable and could produce healthy monkey babies. Transgenic mice have been routinely produced for medical research, and a wide range of transgenic mice are now commercially available. But no transgenic primates had ever been produced. The GFP gene was used in the experiment, which produced the world's first transgenic primate, because it makes it very easy to detect that the gene has been successfully inserted into the egg. Rhesus monkeys do not have any green fluorescent genes in their genome, so if any GFP genes were found in the monkeys, that would be proof that they were transgenic monkeys. Gerald Schatten and Anthony Chan, who supervised the project, had hoped that the GFP gene not only would be present in the DNA of their baby monkeys but also would be expressed so that the transgenic rhesus monkey would have some glowing body parts. The experiment was a partial success—the GFP gene was found in ANDi's cells that were sampled from his umbilical cord, urine, hair, and blood, making him the first transgenic primate reported in the literature. However, no GFP was observed in ANDi, and none of his body parts fluoresced. For transgenic primates to be useful, the inserted gene has to be expressed, and GFP has to be made. Perhaps GFP is being produced in areas of ANDi that are not visible, or its production is delayed—this has been observed in cattle.

RNA coding for the GFP protein has been found in ANDi, indicating that the gene is active. However, no GFP was found. We haven't come across RNA yet. What is the significance of RNA, and why were Gerald Schatten and his colleagues looking for it in ANDi's cells? In order to understand the role of RNA in making GFP, we have to go back and look at cells.

There is some debate about how many cells there are in a human body, though ten thousand trillion (10×10^{15}) is a number often used. Each of these cells has more components than a typical passenger plane.

An average human cell is about twenty microns wide, which is about one hundredth the width of a human hair. They are so small we can't see them with the naked eye, yet they contain thousands of organelles, like mitochondria, ribosomes, and nuclei, as well as millions and millions of molecules. If you could take one of your cells and expand it so that it were half a mile across, you would see molecules the size of basketballs and proteins the size of cars flying around at the speed of bullets.[3] This is not the type of place you would want to store the most important document you have; it would be ripped to shreds within seconds. Cells have evolved a safe place to store the most important documents—the nucleus is where the DNA is kept, protected from most very reactive molecules, like radicals and oxidants, that are flying around the cell. Even in the safety of the nucleus the DNA is damaged ten thousand times a day, which would be catastrophic if there weren't proteins whose sole function is to find the damaged sites and fix them.

In 1869 Johann Friedrich Miescher, a Swiss researcher working at the University of Tuebingen in Germany, was the first to see DNA under a microscope. He was looking at some pus cells obtained from surgical bandages when he saw some material that he had never seen before and had never heard about. The material was located in the nucleus, so he called it *nuclein*. Meischer even suggested that the material was responsible for heredity, an idea so far ahead of its time that it was totally ignored. Of course, he was right; the DNA in the cell nucleus contains all the hereditary information, the recipes to make all the proteins needed in the human body.[4]

Making proteins is a very messy and dangerous business and cannot be done in the nucleus. Instead, the recipe for the protein is photocopied in the nucleus and the instructions are sent to a part of the cell called the *ribosome*, where the proteins can be made without any danger of damaging the DNA. Obviously, there aren't really any photocopying machines in the nucleus, but there are proteins that act like photocopiers. They are responsible for unwinding the DNA double helix and forming an RNA strand opposite one of the DNA strands. This RNA strand has all the information to make the protein, and it acts as a messenger to take all the information of the gene encoded in the DNA from the safety of the nucleus to the protein factory, the ribosome. Sydney Brenner, mentioned earlier, was one of the first scientists to recognize that genetic information was being carried out of the nucleus by a short-lived material called messenger RNA (mRNA).

Although Schatten and his coworkers didn't find any GFP in ANDi, they did find RNA carrying the instructions for GFP in some of ANDi's cells. This means not only that ANDi contained the GFP gene but also that the cellular machinery in the cells was copying the GFP gene and sending the RNA out of the nucleus.

Why did Schatten and Chan go to all the effort to produce a transgenic monkey, and why was their research funded? As noted earlier, it was not to create a fluorescent monkey; instead, the experiment was meant to show that it was possible to insert foreign genes into monkeys. If one can insert GFP genes into a monkey, then one could also insert genes that increase the monkey's risk of contracting diseases such as diabetes, Alzheimer's, and Parkinson's. These primates would be better human models than mice would be, because cognitive decline cannot be studied in animals that do not have the mental complexity of humans.

The process of producing ANDi was a tricky one. It consisted of two steps. First, the GFP gene had to be inserted into the egg, which then, in the second step, had to undergo in vitro fertilization. The GFP gene was inserted into the egg using a disabled virus that was injected into a batch of monkey eggs. In vitro fertilization of human eggs is very common and routine; however, that of monkeys is still in its infancy. This is exacerbated by the fact that it is difficult to collect monkey eggs and that primatologists are not able to artificially control a monkey's reproductive cycle.[5]

Two hundred twenty-four monkey eggs were modified by noninfectious viral insertion of the GFP gene. The transgenic eggs were fertilized in test tubes before being implanted into surrogate mothers. Only forty embryos were formed. Five of the implantations were successful, and three healthy monkeys were born: two had no GFP genes, and one, ANDi, had the GFP gene, but he did not fluoresce. There was, however, a pair of miscarried twins that did have glowing fingernails.[6] Twins are very rare in monkeys; perhaps that is why they were miscarried.

Chan was satisfied with the results. "We have shown that it can be done. Primates can tell you more about what affects humans than mice ever will."[7] One of the problems with using mice as human model organisms is that mice do not get human diseases when genes for these diseases are introduced into them. So in order to investigate more complicated ailments, such as AIDS, cancer, and Alzheimer's, a model closer to humans is needed. ANDi is the first step in that direction.

Gerald Schatten was well known in the scientific community before producing the first genetically modified monkey; however, ANDi brought him a fair deal of public notoriety. As expected this ranged from highly enthusiastic responses from patients' groups seeking faster cures to diseases and from investors in biotechnology, to fear and apprehension from environmentalists, animal rights associations, and church groups.[8] On January 13, 2001, a day after the paper describing ANDi was published in *Science*, the comedy show *Saturday Night Live* reported, "Scientists have fused the DNA of a monkey with that of a jellyfish so as to make it glow green. The results will be reported in the New England Journal of Evil!" In a BBC report about ANDi, the first genetically modified monkey, the spokeswoman of the British Union for the Abolition of Vivisection, Wendy Higgins, said, "This is just the start. Now we are talking about small numbers of animals and gene markers, but what will happen in the future is that scientists will either add or knock out genes in primates to see what happens to them. The end result is terrible suffering. It's bad enough using rodents, but for scientists to play God with primate genes is morally abhorrent."[9]

Schatten's response: "We wouldn't want to make a monkey that carries a disease unless we knew there was a cure right in front of us. Our goal isn't to make sick monkeys. Our goal is to eradicate disease."[10] Schatten and the authors of the *Science* paper announcing the birth of the first transgenic primate have called for a discussion "to solicit comments and then chart directives on the best course of contribution of genetically modified and/or identical nonhuman primates for accelerating cures and discoveries in molecular medicine."[11]

Mice are the most commonly used human model systems in medical research; however, they do have some drawbacks. They have different physiologies and are much smaller than humans. This means many methods of sampling and physiological analysis are not possible on animals the size of a mouse. If larger model systems are required, cats, dogs, or farm animals are used. Primates are used only when the disease being examined requires the use of an evolutionarily and cognitively advanced species.

ANDi has a jellyfish gene, which is quite amazing, but it doesn't prove that a human gene can be inserted into a monkey and that the gene would be expressed in primates, giving them human diseases. However, the chances are very good that a human gene can be expressed in monkeys because our species are so closely related. The great apes are even more closely related to humans than monkeys. About five million years ago, humans and the great apes shared a common ancestor. As a consequence, their genetic information is very similar to ours. Ninety-eight percent of the DNA in humans and chimpanzees is identical. This means that chimpanzees are more like humans than gorillas are. If chimps and humans are so genetically similar, why are they so anatomically different? How can fewer than five hundred out of thirty thousand genes be responsible for difference in appearance and intelligence between, say, Albert Einstein and a chimp? Some of these questions were answered in the early 1980s when fruit fly geneticists found a set of genes that were responsible for telling it how to put together its body and when and where to make the wings, legs, and eyes. Their excitement soon turned to confusion when mouse geneticists found virtually the same set of genes present in mice. They had the same function and were located in the same position as the *Hox* genes in fruit flies. This would imply that the building plan of all animals originated from a long-extinct ancestor that roamed the earth 600 million years ago.

How could the same genes be the architectural plans for two so different species? The *Hox* genes code for proteins called *transcription factors*. Their job, together with the promoters, is to switch on and off protein production. The promoters are stretches of DNA. Specific transcription factors encoded by the *Hox* genes bind to their promoters. Each gene has a number of promoters that activate protein production if their transcription factors are bound to them. So even if zebras and giraffes have the same *Hox* genes, they end up with substantially different bodies because although the proteins making up their bodies are the same, they are produced at different times and in different amounts. Vertebrae development in all animals starts at the head and progresses down to the tailbone. In the giraffe, the promoter responsible for expressing proteins that make neck vertebrae remains switched on much longer than in the zebra, thereby creating a much longer neck.

In 1999 more than sixty-one hundred Americans died waiting for organ transplants, and currently more than eighty thousand patients are waiting for donated organs.[12] A possible solution to this problem is xeno-transplantation, that is, the transplantation of tissues from one species to another. Primates have occasionally been used as a source of donor tissues and organs for humans because human bodies are less likely to reject tissues from primates such as baboons, which are more physiologically similar to humans than other species are. A large amount of research is also being conducted to investigate the possibility of using pigs as donors of organs and tissues. The advantages of using pigs are that they are easily farmed; they give birth to large litters; their organs match the clinical requirements; and they have developmental and immune systems similar to humans. Furthermore, ethical concerns about using pigs as organ donors are mitigated by the fact that annually over ninety-five million pigs are slaughtered as a food source in the United States alone.[13] Before pig organs can regularly be used in humans, the problem of rejection has to be overcome. Humans and monkeys possess natural antipig antibodies, which result in hyperacute rejections when organs from swine are transferred to primates. In order to overcome problems of rejection (after all, which self-respecting human circulatory systems would like to be run by a pig heart), transgenic pigs are being produced that will reduce rejection rates. To date none of the approaches that have been studied have been completely effective. Despite these drawbacks, pigs are used in some areas where the rejection problem has been overcome, such as heart valve and skin replacements.

Transgenic pigs are also used in agriculture. Sows have been generated that produce bovine alpha-lactalbumin in their milk. This protein is normally found only in cows, but the gene for bovine alpha-lactalbumin can be inserted in pigs. Piglets drinking their mother's milk that has been supplemented with proteins produced from this cow gene inserted exhibit increased growth rates. Randy Prather, a professor of reproductive biotechnology at the University of Missouri, Columbia, is one of the researchers in the forefront of the creation of transgenic swine for medicine. He often uses GFP and its yellow mutant YFP as a marker to show that foreign genes can be expressed in transgenic swine. Figure 11 in the photo insert shows two pigs—the one on the right is a regular piglet, a little cleaner than your typical piglet, but no different from the piglets you find on a hog farm. The one on the left is clearly very different from any pig we have ever seen before. It is a transgenic EGFP cloned pig created by Professor Prather. In my opinion, this one photo is more effective at showing what can be done with current genetic techniques than anything I can describe or any other photo I have seen.

It always amazes me that Alba, the GFP bunny, and ANDi, the transgenic rhesus monkey, hogged so much publicity, but the EGFP piglet got virtually none. As was the case with ANDi, the yellow piglet was not created without a good scientific reason. It was formed to show that it is possible to produce a transgenic clone. In the words of Prather, "These animals prove that we can make genetic modifications to express desired traits. For xenotransplantation, this is a large step because it means it's possible to change the genetic makeup of the cells to prevent the body's rejection of transplanted organs."[14] *Sky News* summed up the research: "Scientists have developed the first pig with a fluorescent yellow snout and trotters using jellyfish DNA. Researchers in the US say the work is a step towards growing animal organs for transplants—which could save thousands of human lives. But opponents have said the work is a freak show and a perversion of science."[15]

The piglet was cloned by using a process called "nuclear transfer," the same method that was used to clone Dolly the sheep. But before the piglet could be cloned, the EGFP gene had to be inserted into its genome. This was done by removing some cells from a pig, growing them in the laboratory, and adding the EGFP gene to the genes responsible for producing the tissues in the hooves and snouts. Now that the DNA in the nuclei of these cells contained all the information required to make an identical copy of the original pig, with the addition of some genetically altered GFP, they were removed from the cells by using a very thin hollow needle. Then, they were each inserted into the nucleus of mature unfertilized eggs, called *oocytes*, from which the nucleus had been removed. One hundred ninety-one EGFP nuclear transfer embryos were transferred to three surrogate mothers. Only one maintained the pregnancy, and she produced five piglets. Four of the resulting piglets had yellow fluorescent proteins in their hooves and snout. (See figure 11 in the photo insert.) They were all identical to the pig that donated its cells at the beginning of the experiment, but they had a mutated jellyfish gene expressing EGFP in their snouts and trotters.

At this point, you might be asking yourself why Prather created a transgenic clone and not just a transgenic pig. What is the advantage of using a cloned pig? It turns out that the only way to produce a genetic knockout, as is needed for xenotransplantation pigs, is by creating a cloned pig.

How close are we to being able to transplant major organs from pigs to humans? I would say we still have a while to go. At the October 2002 International Xenotransplantation Congress in Glasgow, David Sachs of Massachusetts General Hospital presented his research on transplanting humanized pig kidneys harvested from Randall Prather's cloned pigs into baboons. The pigs were genetically engineered so that they lacked a key sugar molecule that is normally responsible for the human and baboon immune systems rejecting pig organs. The gene has been "knocked out." Baboons with kidneys from the genetically modified pigs survived for up to eighty-one days, while those with unmodified pig kidneys lived for only thirty days. Eighty-one days is the longest that researchers have extended a baboon's life with pig organs.[16] Clearly, we still have quite a way to go.

Particularly as there are still two major hurdles, besides rejection, that have to be overcome. One major problem with transplanting pig organs to humans is the danger of transferring pig endogenous retroviruses (PERVs) into human patients. This raises the possibility of another retrovirus pandemic, like AIDS, if the viruses mutate to adapt to their human hosts.[17] The other is the much less defined and more complicated moral question, whether it is acceptable to produce genetically modified animals so that we can harvest their organs. There is no easy answer to this highly charged question.

As an aside, before describing some other uses of glowing genes, I want to update you on Dolly the sheep. PPL Therapeutics, the company that created the first cloned mammal—Dolly the sheep—decided to sell all its assets and close down in September 2003, seven years after Dolly was cloned. The company was unable to profitably produce therapeutic proteins in transgenic animals. Dolly died in 2003 and is stuffed and on display at the Royal Museum in Edinburgh, Scotland.

Figure 1. Model swimming in bioluminescent bay, Vieques Island, Puerto Rico. (*Courtesy of Doug Myerscough*)

Figure 2. The firefly. (*Courtesy of J. E. Lloyd, University of Florida*)

Figure 3. *Aequorea victoria. (Courtesy of Osamu Shimomura)*

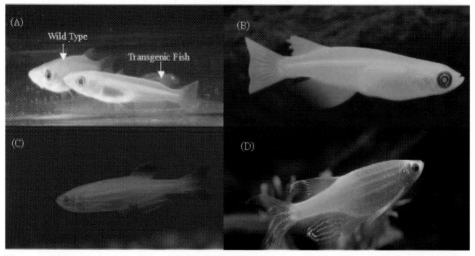

Figure 4. Transgenic zebrafish with (B) GFP, (C) DsRed, and (D) GFP and DsRed genes. The fish are sold in Taiwanese pet shops, where they are known as "Night Pearls." *(Courtesy of H. J. Tsai, National Taiwan University)*

Figure 5. A transgenic GFP daisy *(left)* and a normal daisy irriadited with blue light. *(Courtesy of A. Mercuri, Instituto Sperimentale per la Floricoltura, San Remo, Italy)*

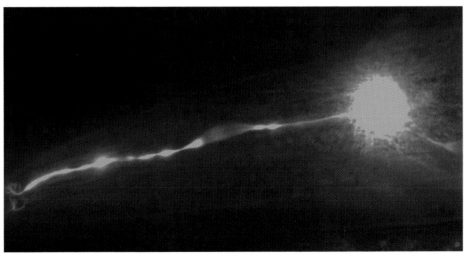

Figure 6. Transgenic *C. elegans* with GFP-labeled odorant receptors. As soon as the round-worm is exposed to diacetyl, it produces the GFP-labeled receptors. *(Courtesy of Tali Melkman and Piali Sengupta, Brandies University)*

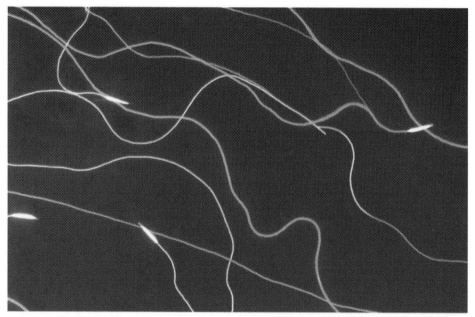

Figure 7. Differentially labeled sperm, dissected from the seminal receptacle of a single doubly inseminated fruit fly female. The green fluorescent tails of the GFP sperm are easily distinguished. *(Photo by Cathy Fernandez and Jerry Coyne. Reprinted with the permission of* Nature.*)*

Figure 8. Alba, the transgenic GFP bunny. *(Courtesy of Eduardo Kac, Julia Friedman Gallery)*

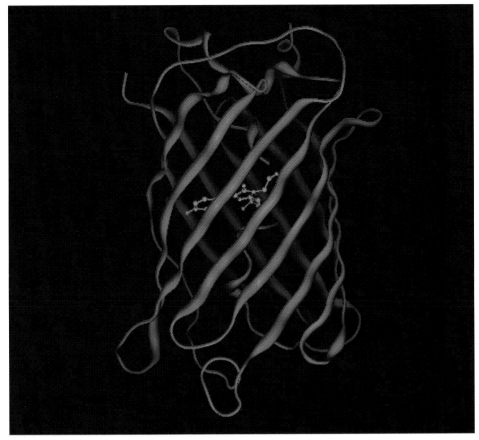

Figure 9. Barrel structure of GFP. The chromophore in the middle of the barrel is responsible for the fluorescent properties of the protein. *(Image created by Marc Zimmer)*

Figure 10. The large star colony shown is not bioluminescent; however, it is fluorescent. Under normal light, it looks quite dull, but it fluoresces with a bright yellow color when irradiated with high-energy ultraviolet light. *(Courtesy of Charles Mazel, NightSea, Andover, MA)*

Figure 11. Two piglets. The one on the right is a regular piglet, a little cleaner than your typical piglet, but no different from the piglets found on a hog farm. The one on the left is clearly very different to any pig we have ever seen before—it is a transgenic YFP cloned pig. *(Courtesy of Missouri University Extension and Agricultural Information)*

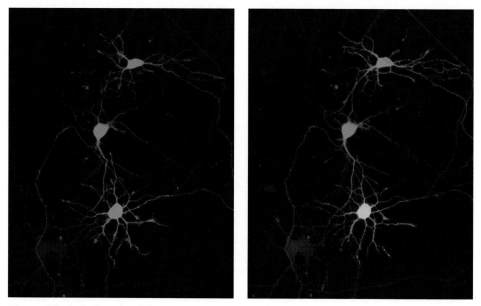

Figure 12. *Left:* Image of three neurons one week after adding the Kaede gene to them. *Right:* Same neurons after illuminating the top neuron for 0.5 seconds and the bottom one for 0.25 seconds. *(Courtesy of Atsushi Miyawaki)*

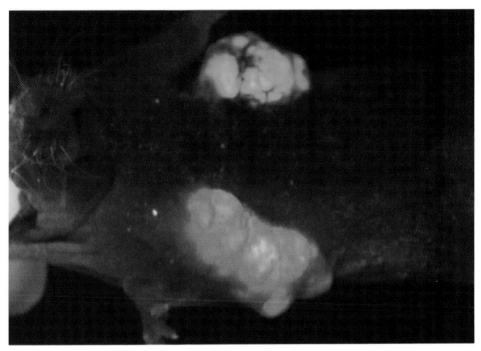

Figure 13. Nude mouse with human breast cancer cells expressing green fluorescent protein *(top)* and red fluorescent protein. *(Courtesy of AntiCancer Inc.)*

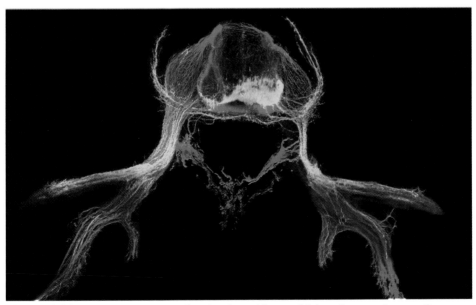

Figure 14. Mouse motor neurons when implanted into a chick embryo and grown from embryonic stem cells extended axons from the spine into the limbs. The mouse motor neurons were labeled with GFP and the green fluorescence of the resulting axons, which project along the nerve branches, can be seen in the figure. *(Photo by Hynek Wichterle and Thomas Jessell. From* Cell *110, no. 3 [2002]: 385–97. Reprinted by permission.)*

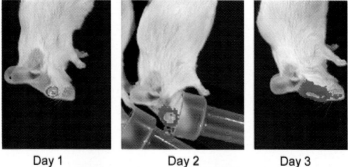

Day 1   Day 2   Day 3

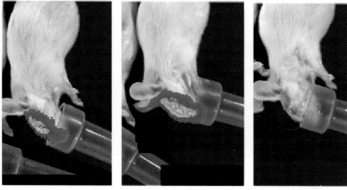

Day 4   Day 5   Day 9

Figure 15. Time course of the spread of the luciferase modified HSV-1 infection, after the cornea of the mouse shown was infected with two million viruses on day 1. The infection peaks after five days and has disappeared by the ninth day. The actual light emitted by the lucifer-ase modified viruses is not shown. Instead, the imaging system counted the number of photons released and converted it to a pseudocolor scale for easier visual examination. *(Photo by Gary Luker. From* Journal of Virology *76, no. 23 [2002]: 12149–61. Reprinted by permission.)*

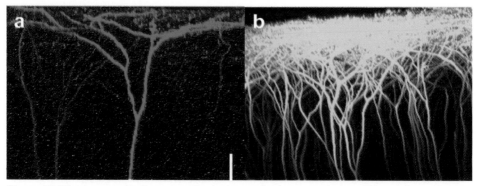

Figure 16. Fluorescent protein expression in the cerebral neurons in two different lines of trans-genic mice. In the one line, GFP is expressed sparsely; in the other, YFP is expressed abun-dantly. Both pictures were taken through a glass window embedded in live mice. *(Courtesy of Brian Chen and Karel Svoboda, Howard Hughes Medical Institute, Cold Spring Habor Lab-oratory. From* HHMI Bulletin *[September 2003]: 27. Reprinted by permission.)*

CHAPTER 12

CAMELEONS, FLIP, FRET, FRAP, AND CAMGAROOS

U p to now, most if not all the applications for glowing genes I described have used the fluorescent proteins as multicolored light-bulbs that are switched on as soon as proteins are made; this way they can be used to monitor the movement of the proteins. Can other techniques be used to extract even more information with fluorescent proteins? Of course, the answer to this question is yes, otherwise I wouldn't be asking it. There are three very useful techniques that I would like to discuss here—they are FLIP, FRAP, and FRET. With names like those, how could I leave them out of this book?

If a fluorescent molecule is zapped with radiation of just the right energy, its fluorescence will be extinguished. FLIP and FRAP both rely on this phenomenon, which is known as *photobleaching*. These techniques are much more effective than the traditional glowing gene methods discussed earlier in showing how proteins move in living cells.

In a FLIP—fluorescence loss in photobleaching—experiment, a small region of the cell is photobleached, and the whole of the cell or organism is repeatedly imaged.[1] The best way of understanding how FLIP works and seeing how effective it is, is to give you an example of how FLIP is used.

A big and important area of biology is developmental biology: How do a sperm and egg join to form an embryo that then develops into a fully formed adult? An interesting question that needs to be answered in the area of cell development is how a cell compartmentalizes itself before it divides into two new cells, each with a different function. We've seen scientists often use model organisms, such as *C. elegans* and the fruit fly, to study such complicated processes. There are many reasons for this—for example, it is not legal to use fluorescently labeled proteins in humans. And "model" organisms are simpler than mammals and are therefore easier to understand. *Caulobacter crescentus* is an ideal model organism for studying compartmentalization. It is a bacteria that undergoes asymmetric cell division to produce two different types of daughter cells called *swarmer* and *stalked* cells. In order to find out when the *Caulobacter crescentus* starts preventing proteins from moving from one side of the bacteria to the other, Harley McAdams and his coworkers at Stanford University School of Medicine developed a FLIP experiment.[2] In the experiment, one end of the bacterium, a single-celled organism, was bleached by tightly focusing a bleaching laser. If the cells are uncompartmentalized, bleaching the cell at one tightly focused position would result in a gradual loss of fluorescence in the whole cell because GFP-labeled proteins will have complete freedom to move all over the cell. This means that they will be zapped by the laser at some point and lose their fluorescence. If proteins cannot freely move around due to compartmentalization, only proteins in the compartment that are being photobleached will lose their fluorescence, while proteins in the other compartment will continue fluorescing. Using fluorescence loss in photobleaching, McAdams showed that in *Caulobacter crescentus* with a 135-minute cell cycle the cells were compartmentalized about eighteen minutes before the cells divided into two daughter cells.[3] Somehow proteins are prevented from moving from one side of the bacteria to the other about eighteen minutes before cell division.

FRAP—fluorescence recovery after photobleaching—is a very similar technique to FLIP. In this method, a region of interest is photobleached with a high-intensity laser, and the migration of fluorescently labeled protein into the bleached area is monitored. FRAP was first developed over twenty years ago to study how molecules "float around" in living cells; this is called diffusion. Lately it has undergone a resurgence due to the discovery of GFP and its analogs and due to the introduction of commercially available laser scanning confocal microscopes that can do FRAP experiments.[4] Computer models are used to interpret the fluorescence recovery after the photobleaching behavior. If the protein of interest, which has been tagged with a fluorescent protein, is free to move around the cell, the fluorescence will recover to its initial prephotobleaching levels, and the shape of the recovery curve can be described by a simple mathematical expression for a single protein. If some of the fluorescently tagged protein is immobilized and cannot move into the photobleached area, then the system will take longer to recover its original fluorescence levels in the photobleached area. A multicomponent diffusion equation (an equation that is based on the fact that there are two or more factors that are responsible for the tagged proteins diffusion behavior) would be able to model the FRAP behavior and provide lots of useful information, such as the percentage of protein that is immobilized and its degree of immobilization.[5]

In the last five years, many questions about protein movement in cells have been answered by using FRAP techniques. In many cases, employing FRAP was the only way to determine how the proteins moved around in the cells studied.

FRAP measurements of GFP-tagged proteins in the cell nucleus have demonstrated that many compartments in the nucleus are not static stable entities. Instead, studies show that the nuclear compartments are collections of proteins that are continuously associating and dissociating to form steady state assemblies. The passage through these compartments is the rate that determines how fast the larger proteins and RNA move within the nucleus.

Many years ago when I was in elementary school, I remember learning that there were things called the *Golgi apparatus* and the *endoplasmic reticulum* that could be found in cells. I was very impressed by these words and thought that I was well on my to becoming a good scientist because I knew about these two impressive-sounding organelles. Traditionally, the contents of the organelles of the secretory pathway, which include the Golgi apparatus, have been thought to be relatively stable resident components. But recent FRAP studies have revealed that the Golgi apparatus is a highly dynamic organelle, the identity of which depends on continuous protein exchange with its surroundings.

In 2002 a GFP-like protein has been found in stony coral, which under normal conditions acts just like GFP, giving off green light when excited. However, it is very sensitive to certain wavelengths of light and changes from green to red when irradiated with these wavelengths.[6] As we shall see in just a bit, this is very useful. So the protein is often used in place of FRAP experiments. The green to red photoconversion was found completely serendipitously. In the paper describing the isolation of the protein and its use, the authors wrote:

> We happened to leave one of the protein aliquots on the laboratory bench overnight. The next day, we found that the protein sample on the bench had turned red, whereas the others that were kept in a paper box remained green. Although the sky had been partly cloudy, the red sample had been exposed to sunlight through the south-facing windows. . . . To verify this serendipitous observation we put a green sample in a cuvette over a UV illuminator emitting 365-nm light and found that the sample turned red within several minutes. . . . At this point the protein was renamed Kaede, which means maple leaf in Japanese.

The color change was irreversible, and the red fluorescence was bright and stable for months. The first experiment scientists conducted with Kaede was a very simple one in which a UV pulse was focused on a spot that was within the cytosolic compartment of a cancerous cervical cell. A red spot of fluorescence was observed after a one-second pulse; the color spread concentrically until it reached the nuclear envelop after 1.2 seconds. After twenty seconds, the red fluorescence had spread over the entire cytosol but was excluded from the nucleus. This shows that although Kaede travels throughout the cytosol, it cannot enter the nucleus. This is not surprising, since the nucleus is the cell's vault where it stores all its important information—the DNA. Satisfied that Kaede could be used in a simple experiment, such as the one described, the researchers decided to use Kaede in a more complex problem.

The brain contains many types of cells, such as glial cells and neurons. It is often crucial to know where one cell or neuron begins and where another ends. In a dense culture of neurons and glial cells, it is very difficult to distinguish between individual cells because they extend long and thin tendrils, called *processes*, that become entangled with those belonging to other cells. When cultured neurons are labeled with fluorescent proteins, they cannot be individually identified.

Adding the gene for Kaede to a neuron culture produced many green neurons that were shown to be entangled with one another and could not be separated. However, individual neurons could be distinguished by focusing a UV pulse on the cytosolic portion of a neuron for ten seconds. This photoconverted the green Kaede to its red form, which spread over the entire cell, including its dendrites and axon, within a few minutes. Figure 12 in the photo insert is an image obtained by confocal microscopy; it shows the single neuron that was irradiated with the 0.5-second UV pulse (red), one that was illuminated for 0.25 seconds (yellow), and the other neighboring neuron in green. Using Kaede it is also possible to label more than two neurons with different colors by UV irradiation with differing duration times.[7]

Fluorescence resonance energy transfer (FRET) is the radiationless transfer of energy from one fluorescent molecule, like GFP, to a different fluorescent molecule, such as blue fluorescent protein. The radiationless transfer of energy strongly depends on the distance between the two fluorescing species. Employing FRET, one can measure interactions between cellular components that are on a scale between 0.000000010 and 0.000000100 meters apart. Closely related to FRET is a technique called BRET, or bioluminescence resonance energy transfer (BRET), which relies on the transfer of energy between a bioluminescent system, such as a luciferase, and a fluorescent molecule, such as GFP. While FLIP and FRAP have extended the use of glowing genes in monitoring the movement of labeled proteins, FRET and BRET have allowed us to "see" something we have never been able to see before—that is, how proteins interact with other proteins and also how proteins can change their shape in living organisms. Let me introduce you to some examples of FRET and BRET.

Roger Tsien uses the analogy of monitoring bracelets for criminals to describe how FRET works. "If, say, the spatial resolution of the bracelets was half a mile, a monitor could tell whether two criminals were in the same vicinity but not whether they might be conspiring. However, if their bracelets interacted with each other in a special way when the criminals got within a few feet of each other, then a signal could be sent back to headquarters. That special signaling is exactly what happens when two different colors of GFP overlap."[8]

In 1994 Marty Chalfie first expressed GFP in *E. coli* and *C. elegans*; not more than two years later, two groups managed to show that FRET can occur between two GFP-like proteins. Both groups used a short stretch of amino acids to link a green fluorescent protein with a blue fluorescent protein. The linker held the two fluorescent proteins close together so that if one of the fluorescent proteins was excited, fluorescence resonance energy transfer to the other fluorescent protein could occur. When the linker was cleaved by some enzyme occurring in the cell, the two fluorescent proteins separated and the FRET diminished. Having shown that FRET between two GFP-like fluorescent proteins can be monitored and that it does depend on the distance between the two fluorescent proteins, many research groups, particularly that of Roger Tsien, began to use the concept to design FRET systems to investigate interactions between proteins and other significant biological processes.

The ultimate goal of many research groups that are designing FRET systems is to make a probe that can detect protein-protein interactions. This is extremely important and hellishly difficult to perform. The problem is that FRET only occurs when the two labeled proteins interact in such a way that enables the two tagged GFP proteins to come close enough to each other to have the correct orientation for FRET to occur. Perhaps a silly example can demonstrate what I mean. Imagine you would like to label two of your friends Bob and Emily so that you could tell when they become intimate in a very dark room. If you try and label Bob with a green glow stick on the small of his back and Emily with a blue glow stick on the small of her back, then the chances are fairly small that the green and blue fluorescence will interact even if Bob and Emily get close. However, if we tag them with glow sticks in their belly buttons, we are much more likely to see the two fluorescence sticks closely approach each other. The same is true for proteins—the GFP-like proteins have to be attached to the interacting proteins in such a way that they are in proximity when the protein-protein interaction occurs. To date, these drawbacks have limited the application of FRET between GFP-like proteins in the screening for protein-protein interactions on a large scale;

however, it sure has proven very helpful in examining specific protein-protein interactions and even small-scale screening.

In 2000 FRET was used to do a small-scale screen to find out which proteins interact with each other as they travel through a nuclear pore complex. FRET between importin receptor proteins tagged with the cyan variant of GFP and a series of yellow fluorescent tagged nucleoporins was monitored in live yeast cells. If FRET was observed between the importin receptor protein and the nucleoporin, it was assumed that the importin receptor interacted with that specific type of nucleoporin. The study indeed found protein-protein interactions between the importin receptor protein and two types of nucleoporins (that were known from different experiments), and they also discovered a new previously unknown importin receptor-nucleoporin interaction that had not been observed before.[9]

Two proteins called Bcl-2 and Bax play a vital role in programmed cell death (apoptosis) as well as in the initiation and progression of human cancer. By labeling Bcl-2 with GFP and Bax with a blue fluorescent protein mutant of GFP, researchers were able to use FRET to show that there was a direct interaction between the two proteins in mitochondria.[10]

Earlier we saw that when aequorin is isolated from the jellyfish and is separated from GFP, it emits blue light, but when they are associated in vivo, green light is given off. This is a natural example of biolumines-cence resonance energy transfer (BRET): the bioluminescent energy of aequorin is transferred to GFP. The same concept can be used to examine protein-protein interactions in the laboratory. One of the first BRET experiments performed by scientists examined the interactions between two proteins involved in the circadian clock of cyanobacteria. One of the proteins was labeled with luciferase and other with GFP.[11]

Light is crucial to most species living on the surface of our planet. In plants it controls growth and circadian responses. There are at least two major types of photoreceptors that detect the presence of light and pass that message onto the appropriate cellular machinery. *Cryptochromes* are blue/ultraviolet light receptors, while *phytochromes* are red light–absorbing receptors. Researchers at the Department of Cell Biology and National Science Foundation Center of Biological Timing at the Scripps Research Institute have labeled a phytochromic protein with DsRed and a cryptochromic protein with GFP to see whether the two proteins interact. By using FRET microscopy, they were able to show that the two proteins interacted in a light-dependent fashion in something called nuclear speckles and that the interaction controls the flowering time and circadian period.[12]

One of the most essential uses of FRET has been in the design of new probes to measure calcium levels in living organisms. Changes in free calcium concentrations control a variety of cellular functions such as contraction, secretion, and gene expression. The study of the role of calcium in cell physiopathology requires the ability to monitor the dynamics of its concentration in living cells, where free calcium ion concentrations can vary by a factor of a thousand. Aequorin, which we first discussed in chapter 4, is a fairly useful calcium detector; however, it has some drawbacks. For almost thirty years, the only source of aequorin was the jellyfish. Extraction and purification were tedious and had to be done in complete absence of calcium. Using aequorin as a calcium probe meant hard work. When Milt Cormier and Douglas Prasher succeeded in cloning aequorin in 1985, they certainly simplified the process.[13] However, there are still some problems with aequorin—it requires the addition of coelenterazine, and its fluorescence is not very bright. Furthermore, aequorin is irreversibly consumed by calcium and therefore cannot be used in long-term studies that occur in calcium-rich environments.[14]

Three new types of calcium sensors have been created that overcome the problems exhibited by aequorin. They all use GFP or GFP analogs together with a protein called *calmodulin*, which changes its shape when it binds calcium ions. The three types of calcium probes are called *Cameleons*, *camgaroos*, and *pericams*.

Cameleons have been devised by Roger Tsien and others. They are based on fluorescence resonance energy transfer between two fluorescent molecules that are linked by a short stretch of calmodulin, a protein that changes its shape in the presence of calcium. In the absence of calcium, the two fluorescent proteins are well separated. However, upon binding with calcium, the two green fluorescent protein mutants are brought closer together. The more calcium that is present, the closer the two fluorescent molecules come to each other, and the more FRET occurs.[15] These calcium sensors are called Cameleons because they change color and have a long tongue (calmodulin) that retracts and extends in and out of its mouth when it binds and releases calcium. Unlike their namesakes, Cameleons are spelled without an H—that's because the chemical symbol for calcium is Ca. Since the first Cameleons were created, they have undergone numerous changes to make them less sensitive to pH. This allows them to form more efficiently at 37°C (98°F) and to make them more sensitive to calcium. The early Cameleons used a green fluorescent protein (GFP) and a blue fluorescent protein (BFP) as the two fluorescent molecules, while the newer, more effective Cameleons use an enhanced cyan-emitting mutant of green fluorescent protein and the yellow-emitting mutant. One of the main advantages of Cameleons is that they are genetically encoded indicators. This means that the gene for the Cameleon, which is a combination of the genes for the two fluorescent molecules and the calcium-binding protein, calmodulin, can be expressed in just the right place, where the calcium concentration needs to be measured without using any invasive methods that would physically place a probe in the desired location. However, despite all the efforts at improving Cameleons, they are still far from perfect. One of the main problems is that Cameleons, which are composed of two GFP-like molecules and a calcium binding protein, are fairly large and complex, which can in some cases significantly impair their targeting efficiency. Therefore, alternative calcium sensors have been developed that are based on a single GFP.[16]

In both camgaroos and pericams, the binding of calcium ions to a calmodulin protein that is fused to a single GFP alters the fluorescent behavior of that GFP's chromophore. There is a site in GFP where relatively long stretches of amino acids can be inserted into GFP without impairing the ability of GFP to fold into its characteristic cylindrical shape and form its chromophore. In camgaroos, calmodulin has been inserted at this site, which is located between the 145th and 146th amino acids. The fluorescence of the yellow mutant of GFP that is used in camgaroos increases when it binds calcium. The calcium probe is named after the Australian marsupial because it is derived from a yellow fluorescent protein that carries a calcium-binding protein (cam) in its pouch.

Unlike Cameleons and camgaroos, pericams were not derived by Roger Tsien's laboratories. However, they are based on an amazing piece of protein origami that was carried out in Tsien's laboratory. He found out that you could create at least ten different fluorescent mutants of GFP in which the first and the last amino acid of the protein were joined together with a linker made up of five amino acids, and the chain was cut at another point. These so-called circularly permuted GFPs were all created by manipulating the DNA that codes for the GFP. It is a bit like starting a cooking recipe halfway through the process, cycling through to the beginning when one reaches the end of the recipe, and finding that the recipe still works. The circularly permuted GFPs have been shown to be highly sensitive to environmental changes and have been used to create a new calcium sensor called a pericam.[17]

Earlier I introduced the grandfather, Osamu Shimomura, and the fathers, Douglas Prasher and Marty Chalfie, of the GFP revolution, and I have been describing the growth of the GFP revolution ever since. Roger Tsien, whose name appears throughout this book and who has been responsible for a large amount of GFP's development, should probably be considered to be GFP's teacher. In an interview with the *Howard Hughes Medical Institute Bulletin*, Tsien modestly claims that he played only a minor role in tuning up GFP for biotechnology—he is conscious that GFP is a naturally occurring protein. "In a way it's like somebody who turns an obscure novel into a popular film," Tsien said. "The basic idea didn't come from us, but we maybe helped lots of people appreciate the stuff that wasn't quite as easy to appreciate in its original form."[18]

In 2002 a major milestone in the development of the GFP revolution occurred. GFP is a fantastically useful protein because it can monitor when proteins are made, where they go, and which other proteins they interact with. All of this can be done inside a living organism, without disturbing any molecular processes. Prior to 2002, there was no equivalent way of establishing what was happening to all the chemicals that were made and broken down in living organisms. Techniques analogous to the glowing gene technology were urgently required to monitor metabolite levels and movements without disturbing the cells being studied. Metabolites are much smaller than proteins; they are more mobile and their concentrations are more variable than those of proteins. This makes developing an in situ probe for metabolites very difficult. Most techniques available at the time were not able to measure metabolite changes in real time or account for likely variations in local metabolite concentrations at the cellular level. In 2002 Wolf Frommer and his colleagues at the Center for the Molecular

Biology of Plants at the Eberhard Karls University in Tuebingen, Germany, created a probe that could be used to measure the sugar maltose in a variety of solutions and in yeast.[19] More importantly, the maltose probe was just a "proof of concept" experiment to show that other probes could be developed in a similar way that would be able to measure the concentration of other metabolites in other organisms. The probe was based on a bacterial periplasmic binding protein. There are a large number of these types of proteins; each type binds a specific compound with a high affinity. The compound binding the protein is often called the *substrate*. Many bacterial periplasmic binding proteins have been well characterized; most of them consist of two lobes that close, in a Venus flytrap–like movement, upon binding their substrate. Frommer and his group chose a bacterial periplasmic binding protein that binds exclusively to maltose and maltooligosaccharides (long chain sugars), and he tagged a cyan GFP mutant on one end of the protein and a yellow GFP mutant on the other side. When the bacterial periplasmic binding protein in the probe bound maltose, it snapped close like a Venus flytrap and brought the two GFP mutants closer to each other, inducing fluorescence resonance energy transfer. Frommer had created a FRET probe that could detect maltose.

To test whether the probe could be used for the rapid analysis of complex solutions, Frommer and his colleagues used my favorite complex solution, beer. The FRET probe was able to determine the amounts of maltose and maltooligosaccarides in some good German beers as well as the typically used laboratory methods. Beer is great, but it is not a living organism, so they decided to test their maltose probes on living yeast cells. They chose yeast because it was easy to control the amount of maltose in the yeast. The probe was expressed in yeast that was specially grown with ethanol as its only carbon source, so that it had little to no maltose. Upon addition of maltose to the yeast, fluorescence resonance energy transfer (FRET) was observed. The maltose probe in the inebriated yeast was able to detect the influx of maltose. A lot of work is still required to perfect the maltose probe, but the system has an enormous potential, especially since there are a large number of bacterial periplasmic binding proteins that selectively bind very important small molecules, including neurotransmitters, vitamins, nitrates, phosphates, and metal ions.[20]

Since the original paper on the maltose sensor was published in 2002, two new publications from Frommer's group have appeared describing the development and use of two new similar FRET-based metabolite sensors. Sensors for both ribose and glucose were constructed and successfully tested in African monkey cells.[21]

Just as babies take some time before they can walk or write a coherent sentence, tests using FRET metabolite probes will take some time to mature and perfect. They are just taking their first steps, but soon they will improve and be more effectively running.

CHAPTER 13

CANCER

According to the American Cancer Society, half the men and one-third of the women in the United States of America will develop cancer in their lifetimes. More than one million Americans are diagnosed with cancer each year, and about 556,000 die from the disease. We have made huge strides in treating cancer, but at the same time people are living longer and their chances of getting cancer are increasing. Before exploring how glowing genes have been used to study cancer, and in at least one case how they have been used to try to kill cancer cells, I would like to tell you a bit about cancer.

Cancer is not a disease—it is a very large group of diseases that can be very different. They all have a common characteristic though, namely, that abnormal cells develop that divide uncontrollably and can infiltrate and then destroy normal body tissue. Worse, as a cancer develops it changes—the cells mutate and no longer respond to the same drugs and treatment. This makes treating cancer very difficult.

Earlier we noted that all cells, except mature red blood cells, contain DNA with all the genomic information required to make all the proteins needed, exactly at the time and place that they are needed. It is this same

DNA that controls the cell division. When cells multiply, it is a very carefully regulated process, and all the genetic information in the DNA is carefully copied. Cancer cells are different; they grow without any controls and spread to other parts of the body. This can occur very rapidly, or it can take place over the course of many years. It is not uncommon that by the time a cancerous mass is detected, the original cancer cell has been dividing for more than five years, and there are a billion cancerous cells present. Although most cancers form tumors, it is not true that all tumors are cancerous. Tumors are abnormal growths of tissue that result from an uncontrolled, progressive multiplication of cells that serves no physiological function.

When abnormal cells are formed in the body, our immune systems often respond by sending out chemical "hit squads" to eliminate them and prevent them from causing diseases. Cancers are able to sneak past our immune systems, especially when they have been weakened by exposure to chemicals or diseases like AIDS. Once the cancer cells have evaded the immune system, they start replicating uncontrollably, spread, and invade and destroy local tissue. This can often have dire consequences; for example, cancer in the brain interferes with its normal functioning and can result in seizures, paralysis, and death. Cancer cells can also upset the chemical balance of their surroundings. Some lung cancers release chemicals that interfere with the control of the calcium concentration and thereby affect nerves and muscles, causing dizziness and weakness.

Aequorin can be used as a calcium probe, and it and other GFP-based probes, such as pericams and camgaroos, have been used to examine changes in the calcium concentration associated with lung cancers in mice.

Regular screening for many cancers, especially cervical, breast, and colon cancers, has been very successful at reducing the fatalities due to these cancers. However, it is impossible to find all cancers in a medical checkup, especially when they are located deep in the body. Cancerous tumors can be diagnosed only by examining excised tumors under a microscope. Computerized tomography (CT) and mammograms can be used to indicate the presence of a tumorous growth. But a biopsy is needed so that a microscope can be used to examine the tumor cells themselves to see whether they are cancerous.

Furthermore, there currently isn't a single technology or an exam capable of detecting a cancerous tumor until the cancer reaches a certain size. This is one of the great difficulties in doing cancer research. How can you observe when a cancerous growth starts when there are only a limited number of cancerous cells, and how can one follow where the cancerous cells go when they metastasize (the medical term for the spread of the cancerous cells)?

Ten years ago, the majority of medical researchers who were attempting to examine cancerous growths in model organisms such as mice would infect many mice with cancerous cells and then periodically sacrifice them in order to perform autopsies. This method was far from perfect because many laboratory animals had to be killed, and it was very difficult to distinguish between environmental and medicinal effects and those caused by the fact that cancerous cells grow and react differently in different individual lab animals. However, this was the only way to observe the spatial and temporal spread of cancerous cells.

Ideally, one would like to be able to monitor abnormal cell spread and their multiplication in live animals. But this is extremely difficult, since most optical methods cannot distinguish between cancerous cells and normal tissue. As a result, they cannot be used to image early-stage tumor growth or metastasis. Intravital videomicroscopy, that is, the use of video cameras that can act as microscopes inside living organs, has been used, but the procedure is too invasive to lend itself to following tumor growth, progression, and internal metastasis. A much more successful idea has been to label the cancerous cells so that they are the source of light. There are two main approaches to doing this—bioluminescence imaging using luciferase and fluorescence imaging with tumor cells tagged with GFP and GFP-like proteins. The advantages of using luciferases are that the light comes from the cancerous growth itself, and the disadvantage is that luciferin has to be injected near the cancerous cells before the luminescence is produced. The advantage of GFP labeling is that it doesn't require injection of any substrates; however, the chromophore in the GFP needs to be excited before it fluoresces, which means that some light, usually UV, has to penetrate through to the cancerous cells.

The first attempts at labeling cancer cells used luciferase as a bioluminescent tag. Most of the work in this area has been done by a husband-and-wife team, Christopher and Pamela Contag, who joined with David

Benaron. It all started in the early 1990s at Stanford University with Pamela Contag, who had managed to insert the luciferase gene into bacteria—salmonella. There are many different kinds of salmonella, and they can cause numerous diseases, ranging from typhoid to food poisoning. Pamela was interested in using her transgenic bacteria to see how salmonella affect the host animal cells that they have infected. Her husband, Christopher, had read an article about detecting light that was passing through living tissue. He knew that David Benaron, a pediatrician and engineer, was using laser beams to probe the inner workings of animals, and this entailed working with extremely sensitive light detectors. So he decided to arrange a meeting to determine whether it would be possible to detect the bacterial bioluminescence in live animal cells. No live animals were used for the first experiment; instead, they used a slab of chicken meat into which they inserted a vial of bioluminescing salmonella. It was a success; Benaron's sensitive photon detectors were able to pick up the light emitted by the bacteria from within the chicken.[1]

After the chicken experiment, live rats were infected with transgenic luciferase–producing salmonella. It was extremely difficult to detect the luciferase bioluminescence within the rats, but it did work. So the Stanford researchers built a highly light-sensitive camera, called a charge-coupled device (CCD) camera, to detect and quantify the luminescence given off by the luciferase. The ultrasensitive CCD cameras can, in addition to measuring light intensity, record the pattern of the light emitted from a sample surface with excellent spatial resolution. Their ideas worked so well and the camera was so efficient that in 1997 they formed a biotech company, Xenogen, to commercialize their technology. Pamela Contag left Stanford University to head Xenogen. The company is based in Alameda, California, and produces transgenic mice and rats that have been engineered to emit light when their luciferase-tagged genes are activated, and Xenogen manufactures sensitive light recorders and software required for imaging. The company has signed licensing agreements with pharmaceutical companies such as AstraZeneca and Bristol-Myers Squibb, which are interested in using the technique for drug discovery. It now employs more than sixty people. Xenogen is currently developing the technology for use with human subjects. (In the next chapter, I will discuss how bioluminescent imaging is being used in drug development and in the monitoring of infectious diseases, but here I will try to limit myself to some of its uses in cancer research.)

Very little research has been done to examine the behavior of cancerous cells when there aren't many cancer cells present. This minimal disease state occurs when the cancer has just started, or after chemotherapy, or after surgery when some cancer cells have not been successfully excised. The reason for this lack of research in both humans and animal models is that there have been no methods available for detecting a small number of tumor cells and monitoring their patterns of growth and distribution over time. This changed in 1999, when Chris Contag and coworkers showed that they could monitor the behavior of a small number of bioluminescently labeled cancer cells in live mice. In order to do this, he took some human cervical cancer cells, modified them by adding the firefly luciferase gene to the cells, and injected them into mice that had severe immunodeficiency. He found that as few as thirty modified human cervical cancer cells could be detected in a live mouse after it had been anesthetized and injected with luciferin. A charge-coupled device camera was used to detect the light emission, which turned out to be proportional to the number of tumor cells present in the mouse over a large range of cell numbers. The researchers were therefore able to monitor the growth of the human tumor cells within the live mice by measuring the increase in luminescence signal intensity over time. Not only were they able to see how the tumor cells were multiplying and spreading, but also they were able to observe the effects of chemotherapy and immunotherapies on the cancerous growths. For example, they observed that the luminescence decreased in mice treated with cisplatin, demonstrating that the drug was killing the human cervical cells.[2] This is the same Bristol-Myers Squibb anticancer drug that was used to cure Lance Armstrong's testicular cancer.

Alnawaz Rehemtulla, an associate professor of radiation oncology at the University of Michigan Medical School, has come up with a very interesting twist of the method pioneered by the Contags and Benaron; he has engineered transgenic luciferase cancer cells that emit light only once the cancer cells have died.[3] Rehemtulla said, "This is the first time anyone has been able to make real-time images of apoptosis—the process of cell death that is so important to so many diseases and treatments. This proves that we can see what's going on at the molecular level while drugs are working, giving results in days or weeks instead of years."[4]

There is always a strict coordination between cell proliferation and programmed cell death in healthy individuals. An imbalance between these two opposing processes results in various diseases such as AIDS and cancer. In order to create his apoptosis probe, Rehemtulla needed to add a "switch" that would turn bioluminescence off while the cell was alive and a switch that would turn it on when the cell was dying. He chose the estrogen receptor as the off switch. It is well known that estrogen receptor turns off the action of attached proteins such as luciferase. Now he needed an on switch. This was a little trickier. Once again Rehemtulla went back to the literature where he found that caspase-3 is an enzyme that is often responsible for initiating apoptosis and that it also cleaves a short protein sequence called DEVD. So Rehemtulla quite ingeniously made a construct that consisted of estrogen receptor linked to luciferase by DEVD. Thus, in the live cell, no bioluminescence was observed because the estrogen receptor was linked to the luciferase and squelched its light-emitting action. Once apoptsosis began, the caspase-3 concentration increased, and it cleaved the DEVD sequence that was holding the estrogen receptor (off switch) and luciferase together. Now that luciferase was free from the estrogen receptor, it began to give off light that could be observed using one of the charge-coupled device cameras developed by the Contags. According to Rehemtulla, the camera is sensitive enough to detect cell death in as few as one hundred cells.[5]

The same two enzymes were also used in one of the first biologically significant examples of how FRET can be used. The DEVD sequence was used to link two different GFP-like proteins. In the absence of apoptosis and caspase-3, the DEVD linker held the two fluorescent molecules in proximity to each other, and fluorescence resonance energy transfer (FRET) occurred. Upon programmed cell death, caspase-3 was produced, the DEVD linker was cleaved, and the two GFP-like proteins moved apart from each other. Consequently, apoptosis, or cell death, resulted in a loss of FRET. This approach was used to screen novel apoptosis-inducing agents.[6]

There seems to be quite a rivalry between groups using luciferase bioluminescence to image tumors and those using GFP fluorescence—

both claim that their method is superior. To date there has been only one study that has tried to compare the two imaging techniques. It was limited to one strain of human breast cancer cells in nude (hairless) mice and gave a slight advantage to luciferase-based imaging.[7] The research was done by a group with significant luciferase experience, however. There is, nonetheless, more than enough room for both methods in cancer research, and both have enormous advantages over traditional methods of trying to follow tumor growth and spread.

Now that we have seen how luciferase can be used, let's now move on to tumor imaging with the fluorescence of GFP-like proteins. In chapter 7, I first introduced AntiCancer Inc. and its work on modifying mouse hair follicles. AntiCancer Inc. is a very interesting company that was started in 1984 by Robert Hoffman. Bob is a biologist with an amazing research career. He has published hundreds of papers and is currently both president of AntiCancer Inc. and a member of the University of California, San Diego, School of Medicine. His parents were both linguists, but he has always been very interested in science. Bob went to the State University of New York at Buffalo, where a professor of his, Phil Miles, reinforced his strong interest in biology and was instrumental in encouraging Bob to carry on his graduate studies in biology at Harvard's medical school.

In the early 1990s, Takashi Chishima (which means "one thousand islands" in Japanese) came to San Diego. He had received his medical degree from Yokohama City University School of Medicine and came to San Diego as part of an exchange program between sister cities. Working with Hoffman, he struggled to find a project that really interested him. After many months of searching for a project that struck his fancy, he noticed the fluorescent *C. elegans* that Marty Chalfie had created on the cover of *Science*.[8] He was immediately intrigued and knew what his project would be—he was going to create a stable cell line of human cancers that expressed GFP. The project really excited him, and he worked very hard making hundreds of constructs placed in dishes spread all over the laboratory. The project was a success, and Takashi managed to create GFP-expressing cancer cells. When these cells were implanted orthotopically in nude mice, their presence could be detected by their green fluorescence. (*Orthotopically* means simply that the cancer cells were implanted in the "correct organ"; for example, lung cancer cells would be implanted in the mouse's lungs and breast cancer cells in its breast.) Nude mice were chosen as model organisms since they have very weak immune systems and don't reject foreign cells—this means that human cancers could be implanted in nude mice. When Chishima and Hoffman published their results in 1997, they reported that they could implant GFP-expressing tumor cells in nude mice and follow the micrometastatic path of the cancer by sacrificing the mouse and looking for fluorescence in the mouse's organs.[9] A couple years later, Meng Yang, Takashi Chishima's successor from Yokohama, was at a conference when he saw an exhibit in which someone was using a blue light with a simple filter system to observe GFP fluorescence in bacteria. The next morning, he snuck a mouse with a GFP-expressing tumor into the exhibition hall and, lo and behold, using the simple filter system, the fluorescence from the tumor was visible in the live mouse. This was a major breakthrough: now tumor growth and metastasis could be followed in live mice without having to sacrifice the animal. Figure 13 in the photo insert shows a live mouse with two tumors, one expressing green fluorescent protein and the other red.

The AntiCancer method is more sensitive and rapid than the traditional methods. It has proven to be particularly good at visualizing metastases in live soft organs and bone. It is equally good in live organisms as it is in the test tube and petri dish.[10] There are numerous advantages to labeling cancer cells with GFP. Only cells that are derived from the initial implanted transgenic GFP cancer cells will fluoresce when illuminated with blue light. This means that the resulting tumors and metastases are selectively imaged with very high intrinsic contrast to other tissue. Since GFP is stable over prolonged periods of time, the growth of tumors, their metastasis, and the effect of antitumor agents can be monitored in just one laboratory animal. Using GFP fluorescence, the folks at AntiCancer have been able to observe internal tumors and metastases in critical organs such as the liver, brain, bone, colon, pancreas, breast, prostate, and lymph nodes. The fluorescence has been bright enough that simple, externally located video equipment could be used to monitor the tumors in mice.[11] However, Hoffman found that in order to get good results, the tumors had to be located near the surface of the animal, mainly because the skin layer absorbed a lot of the fluorescence. To overcome this problem, a minimally invasive, reversible skin flap that can be lifted when measuring the fluorescence was introduced. The sensitivity of detection was dramatically increased so that it is now possible to image metastatic cancer on essentially any organ in a mouse-sized animal.[12] The skin flap "window" has made it possible to directly observe tumor growth and metastasis as well as tumor angiogenesis and gene expression, often down to a single cell level. One of the major advantages of GFP tumor labeling over the equivalent luciferase technology is that the GFP imaging requires just a blue light and filter (about $400), while luciferase photon emission can be detected only with a CCD camera (more than $100,000).

Angiogenesis has became a very "hot" area of cancer research. Many researchers and funding agencies are hoping that angiogenesis inhibitors will be found that can stop the spread of some types of cancerous growths. In order to grow and spread, cancer cells need oxygen and nutrients. They are supplied with these essentials by blood vessels formed by neighboring healthy tissue. This is called *angiogenesis*. Some chemicals stimulate angiogenesis, and others called angiogenesis inhibitors signal the stop of blood vessel production. Researchers are studying several natural and synthetic angiogenesis inhibitors for their potential to halt the growth of cancer by preventing the formation of new blood vessels.

In animal studies, angiogenesis inhibitors successfully stopped new blood vessels from forming, causing some cancers to shrink and die. Unfortunately, initial clinical studies using angiogenesis inhibitors with human subjects have been somewhat disappointing; the research is, however, continuing.

It is very difficult to obtain precise data relating angiogenesis to tumor growth and metastasis if mice need to be killed and their organs inspected for each measurement. For the last three years, researchers at AntiCancer Inc. have been using GFP to study angiogenesis in a way that doesn't require sacrificing animals. Initially implantation and injection of GFP-labeled tumor cells was the only way of visualizing angiogenic blood vessel growth in both space and time.[13] In a relatively noninvasive way, blood vessel development could closely be followed and quantified in real time, allowing researchers to get precise answers to the relation of cancer growth and angiogenesis. The method was used for the rapid evaluation of drugs that might affect the development of these vessels.[14] Unfortunately, only cancerous cells were labeled, and while it was easy to visualize the tumor, it was a lot harder to see the host cells that formed the angiogenic vessels. In late 2003, Robert Hoffman and his colleagues at AntiCancer Inc. found a way to overcome these problems. They implanted red fluorescent protein–expressing tumors into transgenic GFP mice; they make GFP in each and every cell. In this way, the researchers generated lab mice in which all the cells that derived from the tumors were red fluorescent and all the cells originating from the host mouse gave off green fluorescence.[15] If you are interested, the AntiCancer Web site has amazing images of red and green fluorescent tumors, angiogenesis, and transgenic mice (http://www.anticancer.com).

As mentioned before, metastasis is a very important aspect in the spread of cancer. To successfully metastasize, tumor cells have to leave the primary site, invade the local host tissue, enter the circulation, come to rest at a distant site, and proliferate at that distant organ site. Oral squamous carcinoma cells (soft cancerous cells found in the mouth) can invade adjacent tissues and vascular systems early in their tumor progression. Dr. Koh-ichi Nakashiro of the Department of Oral and Maxillofacial Surgery at the Ehime University School of Medicine, Japan, has employed GFP to detect the circulation of oral squamous carcinoma cells in nude mice and has further used the information gained to monitor the circulation of the cells in the blood of human patients.[16] He first detected and visualized circulating cancer cells expressing GFP in the tumor-bearing mice. Using GFP fluorescence as an indicator of oral squamous cell carcinoma metastasis, Nakashiro then tested the utility of two target genes, both of which have been used as markers for circulating cells of other organs. Both were useful in the mouse models and were therefore tried in humans. Circulating cells were detected in the patients; however, there was no clear correlation between the number of circulating cells, quantified by the two markers, and the metastatic status of the patients. This suggests that the process in which the cancer cells enter the circulation, intravasation, is not important for establishment of metastasis; rather, the ability to grow at distant sites maybe more important for the further progression of oral squamous cell carcinoma.

Both the GFP- and luciferase-based methods for following the movement and spread of cancer cells (discussed in this chapter) rely on modifying cancer cells by inserting a glowing gene. This is very useful in studying the behavior of cancer cells in living model organisms, but it is not very practical in the direct treatment of cancer in humans. Aladar Szalay, who has positions at both the University of Würzburg in Germany and the School of Medicine at Loma Linda University in California, has developed a new method for monitoring primary tumors and metastases in animals that might soon be used in humans. The technique is based on the observation that some microorganisms, such as viruses and bacteria, are present in tumor tissue excised from human patients and that they are absent in the rest of the body. Szalay has genetically modified these microorganisms with light-emitting proteins such as firefly luciferase and GFP so that he can visualize their movement from the bloodstream into the tumor region and their replication in solid tumors.[17]

He has discovered that light-emitting bacteria colonized tumors to such an extent that a week after injection, none of the bacteria were found in the blood or internal organs. They were all localized at the tumor. In fact, they were so restricted to the tumor that they were not released into the circulation at all, or at least in sufficient numbers, to be able to colonize other newly implanted tumors.

While most of the GFP-labeled bacteria collected in the center of the tumors, a study of GFP-labeled viruses showed that they aggregated at the periphery of the tumors where the fast-dividing cells are located. This was perhaps because these cells had weakened immune systems. GFP fluorescence from labeled viruses may therefore be used as a marker for identifying tumor margins in order to facilitate precise tumor removal during surgery in the future. This would be an excellent tool for surgeons in knowing exactly what to cut out.

All the methods I have described in this chapter, up to now, have been glowing-gene techniques that were designed to monitor the growth and spread of cancerous cells. These methods are excellent alternatives to conventional techniques. They have become a very important tool in cancer research and in drug testing for new forms of cancer treatment. However, they are not anticancer treatments. In April 2003, a paper describing the use of firefly luciferase to cause cancer cells to self-destruct appeared in the journal *Cancer Research*.[18] This is the first time to my knowledge that glowing genes themselves are the active agent causing the elimination of cancerous tumors. I hope this method will one day be used to treat cancer in human patients. The method is a modification of a treatment currently used called *photodynamic therapy.*

Photodynamic therapy has been around for about twenty years and is used to kill tumors that are located close to the surface of the body. The tumors are treated with photosensitizers and irradiated with light or lasers. Once the photosensitizers have been exposed to light, they produce a toxic form of oxygen called *singlet oxygen* that destroys the cancer cells by rupturing their membranes. A disadvantage of photodynamic therapy that has limited its use to surface tumors is that light and laser beams cannot penetrate very far into the body. However, *Bio*Luminescence *A*ctivated *De*struction of cancer, or BLADe, is a new technique developed by Theodossis Theodossiou, a physicist at the National Medical Laser Centre at University College, London, that has extended the reach of photodynamic therapy to all parts of the body. "It's a very strange hybrid between a Trojan horse and a guided missile system," said Theodossiou when describing BLADe.[19] You can probably guess how Theodossiou wants to use luciferase to kill tumor cells that cannot be reached with external light sources. He wants to selectively insert the luciferase gene into tumor cells and then inject the cells with the photosensitizer and luciferin. In this way, he is hoping that the photodynamic therapy will kill deeply buried cancerous growths without affecting the neighboring cells. The research is only in its initial stages but is quite promising. Initial tests have shown that only one treatment of BLADe was required to kill almost 100 percent of fibroblasts; these are noncancerous cells often used in research to model cancer because they multiply rapidly like cancer cells.[20] John Hothersall, a microbiologist at the Institute of Urology and Nephrology at the University College in London, worked on the project with Theodossiou and says that the beauty of the treatment is that it can be targeted so that it gets only the cancer cells; therefore, the treatment could be repeated as often as required if the cancer returns. Theodossiou and Hothersall have begun trials on mice with prostate cancer, and if those studies go well, they are hoping to begin human trials in 2005.

CHAPTER 14

GLOWING GENES IN MEDICINE

Think of any medical condition, and I am sure you will find that glowing genes have been used to study that disease. We have seen how GFP and luciferase have been used in cancer research. Here we will explore how they have been used in other areas of medicine.[1]

Let's start with stem cells. They are a controversial subject that has appeared a lot in the news lately. Stem cells are unspecialized cells that can give rise to specialized cell types. Before you were an embryo, you were a bunch of stem cells, which at just the right time differentiated into specialized cells like neurons and red blood cells and ended up making your heart, liver, brain, spleen, and belly button. There are two types of stem cells—embryonic stem cells and adult stem cells. Embryonic stem cells have the potential to differentiate into two hundred different types of tissue in the body. Adult stem cells are unspecialized cells that are able only to give rise to the specialized cell types of the tissue in which they are located and perhaps a few others. For example, we can expect liver stem cells to produce cells required in the liver, but not in other tissue.

In the beginning, at the time of conception, the fertilized egg contains dividing cells that will form the placenta and embryo. They are known as *totipotent cells* because they have the potential to form the entire organism. It takes about four days before the totipotent cells start specializing to form a hollow ball that is composed of an outer shell of cells and a cluster of cells known as the inner cell mass. In time, the outer shell will form the placenta, while the inner cell mass will form the embryo. Human embryonic stem cells are obtained by removing the outer shell of cells or by culturing cells from the inner cell mass. Since these human embryonic stem cells can give rise to all types of cells, but cannot form another embryo, they are known as *pluripotent cells*. However, due to their unlimited self-renewal and boundless development potential, human embryonic stem cells have vast promise in both the treatment and study of many important diseases.

Organ regeneration from human embryonic stem cells can not only halt disease progression but also aid in the repair of damaged organs. Investigators are already using human embryonic stem cells to create heart muscle, brain and pancreas islet cells, and blood vessels.[2] Nonetheless, there are some roadblocks that are currently preventing researchers from placing the cells into patients' bodies. One problem is that the cells cannot be grown without using mouse cells, which means that they could be contaminated with mouse proteins. Another problem is that scientists don't yet know how to control the transformations from one cell type to another. The use of embryonic stem cells also presents many ethical and legal challenges. Most human embryonic stem cells are obtained from unused embryos that have been produced by in vitro fertilization or from an already aborted fetus. On August 9, 2001, President George W. Bush passed an executive order to limit federal funding of human embryonic stem cell research to cells derived from one of sixty-four preexisting cell lines.

Transgenic cloned EGFP pigs, such as the one that graces the cover of this book, are currently being used in retinal stem cell research. Humans do not have the capability to regenerate the optic nerve or photoreceptors following injury; consequently, permanent loss of sight often results. The optic nerve is the largest sensory tract of the human central nervous system, and it connects the eye with the visual centers of the brain by way of approximately 1.2 million separate axons from each retina. The organization of these fibers is critical for maintaining an accurate topographic map of the visual world. Attempts to repopulate the retina with grafted neurons have been unsuccessful, in large part because donor cells prefer not to integrate with those of the host. Many groups have shown that embryonic, neonatal, and mature neuroretinal tissue can be grafted into the retina of adult rodent hosts. This tissue can survive, differentiate, and even form apparent connections with the host retina.

Demonstrating functional integration between graft and host, however, has proven to be a daunting task. Mike Young from the Schepens Eye Research Institute at Harvard Medical School and Henry Klassen of the Children's Hospital of Orange County, California, fly into Missouri to collect fetal retinas from one of the EGFP lines that Randall Prather's laboratory has created. They take them home and isolate retinal stem cells. Then they fly to Denmark, where they collaborate with a group that damages the retina of pigs. The transgenic EGFP retinal stem cells are transferred into the retina of the eye in an attempt to repair the damaged retina. Some of the EGFP cells seem to be repairing the retina by forming rods and cones. This is very exciting work for people who have damaged retinas, and the initial results seem promising.

Lou Gehrig's disease, or amyotrophic lateral sclerosis, affects about twenty thousand Americans, with five thousand new cases being diagnosed each year. The disease causes degeneration of the nerve cells in certain regions of the brain and spinal cord that control the voluntary muscles. This leads to loss of control of the limb, mouth, and respiratory muscles. Since the first human embryonic stem cells were isolated in the late 1990s, researchers have been trying to convert them into replacement motor neurons for those damaged by the disease. It is not a simple process because there are hundreds of distinctly different neurons that establish the diversity that is required for the formation of neuronal circuits. Directing stem cells in a specific manner to produce a desired type of neuronal system has proven to be very difficult because the normal developmental pathways that generate most classes of central nervous system neurons are not well known and are difficult to study.

Spinal motor neurons are a type of neuron for which the pathway of development is known. Researchers at Columbia University have engineered embryonic mouse stem cells that express GFP in the motor neurons. The modified stem cells were grown in vitro, and the scientists added two signaling proteins known to differentiate neural cells in live mice. First, retinoic acid was added to stimulate the formation of spinal cord cells from the stem cells, and then a protein with the fantastic name *sonic hedgehog* was added to convert the spinal cells to spinal motor neurons. About 30 percent of the embryonic stem cells developed into motor neurons. But would they work like motor neurons? Dr. Hynek Wichterle, a postdoctoral researcher in Thomas Jessell's Columbia laboratory, tested the motor neurons by inserting them into a chick's spinal neuron cord. They worked—green fluorescent mouse motor neurons grew long axons that connected with the muscles between the ribs. (See figure 14 in the photo insert.) The figure appeared in the August 9, 2002, issue of *Cell*. It was the first indication that someday in the future, embryonic stem cells might be used to repair damage caused by Lou Gehrig's disease.[3] In an interview with a Columbia University publication, Wichterle said, "I was pleasantly surprised at how well the stem cell–derived neurons mimicked the chick's neurons. But these experiments with embryonic stem cell–derived motor neurons are only the first step in our research for a potential treatment for amyotrophic lateral sclerosis. They open the way for subsequent experiments to determine which cells should be introduced into an adult animal with motor neuron degenerative disease."[4]

Columbia University has applied for a patent for its process of rational nerve generation and for tagging specific nerve cell types with GFP. The university hopes the protocol can also be used in therapies for other neurodegnerative disorders, such as Parkinson's disease.[5]

Until recently it has always been accepted that adult stem cells replace damaged cells in local tissues and that they were more restricted in their development potential than embryonic stem cells. However, in the last three years, it has been reported that adult stem cells also help in disaster relief of other, more distant organs.[6] If this turns out to be accurate, it would be enormously important, since it implies that tissue-specific adult stem cells might be used in place of embryonic stem cells, removing the need to collect stem cells from human embryos and thereby overcoming many of the legal and ethical barriers to stem cell therapy. This would mean that scientists might one day be able to take adult stem cells from our skin and get them to differentiate into neural cells that can be used to replace damaged spinal cord cells.

Many researchers had pinned their hopes on blood-producing adult stem cells as replacements for embryonic stem cells because early research had indicated that they could be reprogrammed to make many other types of tissue. Irving Weissman and Amy Wagers's stem cell biology laboratories at Stanford University decided to use GFP technology to find out whether this were true.[7] They tagged a single adult blood-producing stem cell with GFP, then injected it into a mouse whose bone marrow had been destroyed by radiation. Weissman found that within a few weeks, the lone stem cell had repopulated the mouse's immune and blood system, but it didn't create any other type of cells. Out of fifteen million cells that were examined, only eight fluorescent cells were found outside the blood and immune systems—one in the brain and seven in the liver. According to Amy Wagers, the results clearly show that blood-producing adult stem cells will be very useful in combating blood disease but will not be much help in combating diseases affecting other types of cells.[8]

Similarly disappointing results were found in a related study that examined the extent that adult bone marrow–derived cells participated in long-term repair processes after strokes. Stroke is the third-leading cause of death and the leading cause of adult disability; only cardiovascular disease and cancer cause more deaths annually. Bone marrow–derived cells participate in repair mechanisms of diseases caused by a lack of oxygen, such as stroke and heart attack, which has raised the hopes for the use of bone marrow as a source for cell-based therapeutic approaches. Unfortunately, the latest studies with GFP-labeled bone marrow cells six weeks and six months after a heart attack show that no bone marrow–derived cells differentiated into replacement cells for those damaged by the heart attack.[9] Perhaps embryonic stem cells are the only solution.

Skin stem cells play a central role in tumor initiation, as well as in wound repair and gene therapy. However, the study of skin stem cells had been limited by the absence of tests that confirm the presence and the number of skin stem cells in a given amount of tissue. That all changed in September 2003, when Ruby Ghadially published a method for determining stem cell frequency in the epidermis.[10] As you might suspect, it was based on glowing-gene technology.

In order to find out how many skin stem cells are present in the bottom layer of the skin, the basal epidermis, the researchers at the San

Francisco VA Medical Center used a series of mice that had a patch of their skin removed. Skin cells from two donor mice were placed on the patch. The cells of one of the donor mice were labeled with GFP and those of the other weren't. Skin cells have a short life span and are constantly being replaced by new cells formed by the stem cells. A month after the original mouse had been patched up with skin cells from the two donor mice, the researchers looked for the presence of green fluorescent skin cells. If they could be detected, it meant that some GFP-labeled skin stem cells from the donor mouse were producing skin cells in the test mouse; differentiated skin cells from the donor mouse would not have lasted a month. Having confirmed that they had transferred some skin stem cells, the researchers still wanted to find a way of estimating the number of skin stem cells in the basal epidermis. They achieved this by keeping the number of nonlabeled donor cells constant over a number of host mice but varying the number of GFP-labeled donor cells to see how small a sample they could add before no more differentiated green skin cells were visible after a month's time. This process, called *limited dilution*, gave them a ratio of about one stem cell for every ten thousand skin cells. If they used fewer than ten thousand skin cells, no fluorescence was visible after a month because no GFP-labeled stem cells were donated. If more than ten thousand skin cells were used, there was usually at least one skin stem cell in the donor cells, which produced green fluorescence in the skin one month later.[11] A similar one-to-ten-thousand ratio between differentiated and stem cells is known to occur in bone marrow.

Knowing the number of stem cells present in tissue and being able to distinguish them from other cells is extremely important. Now that scientists are able to identify skin stem cells, scientists must isolate them so that they can be used as effective carrier cells for gene therapy. They could also be used as stem cells in the treatment of burns and skin wounds in much the same way stem cells isolated from blood are now being used to treat cancer patients whose blood cells are damaged by radiation treatment or chemotherapy.

Diabetes is a very common group of diseases that affect the way the body uses sugar. It is the seventh-most common cause of death in the United States and affects about sixteen million Americans. Normally the amount of sugar allowed into the cell is carefully controlled by the release of insulin. When cells require more energy, insulin is released and unlocks the cell "door" so that sugars can enter the cell and be used as fuel. There are two types of diabetes: type 1 is less common, affecting between 5 to 10 percent of the people with diabetes; it develops when the pancreas makes little or no insulin. Type 2 diabetes occurs in 90 to 95 percent of diabetes patients over the age of twenty. In type 2 diabetes, the cells do not respond to the insulin. In both cases, the cells don't get enough sugar, and it is emitted in the urine. More Americans have diabetes than ever before mainly because they have increased their food consumption by an average of more than two hundred calories per day.

Most of the insulin that controls the sugar uptake is produced in cells called beta cells that are found in the pancreas. GFP and its yellow and cyan mutants have been expressed in pancreatic beta cells in order to get a better understanding of their anatomical and functional loss in the development of diabetes. A line of live mice has been engineered with GFP-tagged beta cells in order to try to find the adult stem cells that give rise to the insulin-producing beta cells. The hope is that one day they might be used in the treatment of diabetes.[12]

And now for something completely different.

Our bodies are continuously exposed to bacterial and viral threats. Fortunately, we have developed defense systems against most of these attacks. We have complex and ingenious immune systems made up of many different type of cells that are spread all over the body, where they find and destroy the invading viruses and bacteria. The immune systems are so complex that we are just beginning to understand how they fight off bacterial infections. It is extremely difficult to follow the interactions between the invading bacteria and individual immune cells in "larger" animal models such as mice. *Caenorhabditis elegans*, our familiar old friend the roundworm, has been very useful to study the response of the immune system to bacterial infections such as *Pseudomonas aeruginosa* and *Salmonella typhimurium*. However, the immune system of *C. elegans* is very different from the complex immune system found in mammals. We humans have cells called *macrophages*, which are large cells that are able to surround and then swallow bacteria. The process of engulfing the bacteria is called *phagocytosis*. The roundworm has no macrophage-like cells that are able to phagocytose bacteria.

Wilbert Bitter from the Department of Medical Microbiology at the Free University in Amsterdam, the Netherlands, decided to use zebra fish in place of *C. elegans* in his studies of the immune system.[13] His reasons for choosing the zebra fish are that its immune system is very similar to that found in humans and that the zebra fish embryos develop externally and are transparent during their development. Furthermore, macrophage-like cells appear in the embryos twenty-five hours after fertilization. This means that bacterial infection can be studied in the zebra fish embryo, which can be observed under a microscope for long periods of time. In contrast, the fully developed active adults wriggle around under the microscope, making it very hard to observe the viral infection. Bitter and his helpers infected twenty-eight-hour-old zebra fish embryos with DsRed-labeled *Salmonella typhimurium*. By using multidimensional digital imaging spectroscopy, they were able to determine the exact locations and fate of the bacteria in the zebra fish for three days after infection.

Typically, high magnifications of four hundred to one thousand times are required to observe bacteria. This makes it extremely difficult, if not impossible, to follow the bacteria in a live host like the zebra fish. Lower resolutions can be used if the bacteria can be made highly fluorescent. Bitter first tried GFP-labeled bacteria in zebra fish, but he wasn't very successful because the zebra fish themselves give off some background green fluorescent radiation. Next, he tried DsRed. The bacteria were highly fluorescent, and the zebra fish gave off no red fluorescence of its own, so the bacteria were easily visible. However, if you think back to the chapter about DsRed, you should remember that one of the draw-backs of DsRed is that it takes about 24 hours to form its chromophore, which means it starts fluorescing only after twenty-four hours. Fortu-nately, in 2002, a DsRed mutant was found that formed its chromophore in an hour or so. Bitter found that this DsRed mutant was just what he needed and that DsRed-labeled bacteria could be observed at much lower magnifications than GFP-labeled ones. In fact, he could distin-guish single fluorescent cells at magnifications of 60X; this allowed him to monitor the fate of each single bacterium and determine the actual site of bacterial replication.

Approximately fifty transgenic DsRed-labeled *Salmonella typhimurium* were injected into twenty-eight-hour-old zebra fish embryos. The fluorescence labeling showed that after their initial sojourn in the bloodstream, the macrophage-like cells captured all the *Salmonella typhimurium*. Some of the bacteria were killed in the macrophage-like cells, but most resisted degradation, and some even started to multiply. A stalemate between the macrophage-like cells and bacteria was observed for the first twenty hours after infection, with macrophage-like cells surviving the bacterial onslaught and managing to keep the bacteria isolated within themselves. After twenty hours, bacterial fluorescence spread to areas where there was a reduced blood flow, and therefore where there were fewer macrophage-like cells. Here, the bacteria formed microcolonies in which they actively replicated before making their way to other areas of the embryo. All zebra fish embryos infected with about fifty wild-type *Salmonella typhimurium* died within forty-eight hours of infection. At the final stages of the infection, the amount of bacterial fluorescence and therefore the number of bacteria increased dramatically.[14]

It is well known that certain mutants of *Salmonella typhimurium* do not harm zebra fish or mammals; they are nonpathogenic. Bitter repeated his experiments with these mutant bacteria to see how the immune system of the zebra fish responded to these bacteria. Why didn't they kill the zebra fish embryos? Hardly any of the wild-type *Salmonella typhimurium* were broken open or lysed while they were in the bloodstream during the first hour of the infection, whereas more than 30 percent of the nonpathogenic mutants were lysed in the first hour. However, injection of the same mutant bacteria in the yolk of the embryo resulted in uncontrolled bacterial proliferation. This showed that the blood of the zebra fish has some lytic activity. As yet no one is quite sure how the blood lyses the mutant bacteria and why it can't lyse the wild-type *Salmonella typhimurium*.[15]

About 80 percent of Americans have been infected by herpes simplex virus type 1 (HSV-1). It usually causes cold sores, also called fever blisters. However, in some cases, it can cause genital herpes, a condition normally brought on by herpes simplex virus type 2. In extreme and unusual cases, it can lead to blinding keratitis. HSV-1 spreads very quickly and easily; often shared eating utensils, razors, and towels are responsible for cold sores being distributed within families and households. The virus can lie dormant in nerve cells located in the skin and can reemerge as an active infection on or near the original site. Exposure to sun, fever, and menstruation can all trigger a recurrence. Common antiviral drugs such as valacyclovir (Valtrex) are used to treat HSV-1 infections.

In order to get a better understanding of viral diseases and how they spread in humans, scientists often examine the spread of the herpes virus in mice. For example, studies of HSV-1 infection of mice are often used to gain insights into viral and host genes that regulate the spread and virulence of viral diseases. Conventional methods for studying HSV-1 infection rely on culturing samples from accessible surfaces and from sacrificing the mice to determine the distribution and concentration of viruses in the mouse. Of course, once the animal is dead, it is impossible to continue monitoring the spread of the infection with time. Prior to the use of glowing genes, the only way to find out how the numbers of HSV-1 change with time was by sacrificing a series of mice at different times and therefore different stages of infection. This would not be a problem if all mice were the same, but it is problematic since they are not all identical, and it is difficult to distinguish between temporal effects and variations between individual mice. The spread of HSV-1 to unexpected anatomic areas might also be missed through traditional methods because the infected tissue was not assayed for virus, as it is too time-consuming and expensive to assay all of the mouse for viral infection.

Researchers from the Department of Ophthalmology and Visual Sciences at the Washington University School of Medicine have used Chris Contag's noninvasive, whole-body imaging techniques to detect HSV-1 in live mice and to provide new insights into disease origination, development, and treatment. Herpes simplex virus type 1 was created with firefly luciferase and sea pansy luciferase as reporter proteins. Viral infection in mouse footpads, stomach cavity, brain, and eyes could be detected by bioluminescence imaging of firefly luciferase produced within the HSV-1 with the expensive high-tech camera developed by Contag. The magnitude of the bioluminescence measured in the living mouse correlated directly with the amount of HSV-1 used for infection. When the mice were treated with valacyclovir, an antiviral drug that prevents HSV-1 from replicating, dose-dependent decreases in bioluminescence were observed. Figure 15 in the photo insert shows the spread of the luciferase modified HSV-1 infection. The cornea of the mice were infected with two million viruses.[16] The pictures also demonstrate the effectiveness of the medicine.

HSV-1 is normally limited to localized infections; however, in newly born babies and in patients with AIDS or other immuno-compromising diseases, it may spread to organs such as the lungs and liver. Once HSV-1 infection spreads to more than one organ, it leads to severe sickness and even death, in spite of antiviral treatment. HSV-1, spread into the central nervous system, can cause potentially fatal encephalitis. In order to find out which components of the host immune system limit HSV-1 infection, the same Washington University School of Medicine researchers also infected mice that were genetically deficient in certain genes known to be vital to the immune system with bioluminescent HSV-1. Interferons are known to be the most significant components of innate immunity to viral infection. There are many types of interferons. In mice without type I interferons, the luciferase bioluminescence showed how HSV-1 spread to lungs, liver, and spleen from both ocular and footpad infections. In mice without type II interferons, no systematic spread of HSV-1 or lethality was observed.[17] Clearly the type I interferons have a much more important role in combating HSV-1 spread than type II interferons. It is, therefore, the type I interferons pharmaceutical companies interested in combating HSV-1 should concentrate their efforts on.

Karel Svoboda at the Cold Spring Harbor Laboratories on Long Island is doing something no one else has ever done before. He is watching how mice think. He doesn't sit and watch a mouse in a maze with a puzzled expression on its face—no, he watches the brain of a mouse react to new experiences. How does he do this? With GFP, of course. Joshua Sanes, a collaborator of his from the Washington University School of Medicine in St. Louis, created a transgenic mouse strain that expresses GFP in some of the neurons in the cortex. Then, together with some of his students and collaborators, Svoboda has replaced sections of the skulls of these transgenic young mice with transparent windows so that they can watch what happens to this region of their cortices, which processes sensory information derived from their whiskers. The mice can live out their entire lives with the windows in place, allowing Svoboda the opportunity to monitor the changes occurring over many weeks. He observed tiny spines along the dendrites rising and receding.[18]

The rate of spine turnover increased as the mice were exposed to new experiences. "The spines probably establish new synapses. If the synapse is a useful one, the spine will stay, if not it will retract. The neurons are constantly exploring alternative arrangements, which probably has something to do with learning," Svoboda said in an interview with *Discover*.[19] Figure 16 in the photo insert shows YFP and GFP labeled cerebral neurons in two lines of transgenically modified mice.

In previous chapters, I have introduced GFP's characteristic barrel shape, shown in figure 9 in the photo insert. It folds into its three-dimensional structure because of its amino acid sequence. Sometimes something can go wrong, and despite GFP having the correct amino acid sequence, it doesn't fold into the barrel shape. It is misfolded, and the chromophore doesn't form; subsequently, the GFP won't fluoresce. When protein misfolding occurs in the brain cells, the consequences are catastrophic. Misfolded proteins are a common characteristic of Parkinson's, Alzheimer's, and Huntington diseases.

Scientists at the Department of Energy's Los Alamos National Laboratory have discovered a new method for rapidly determining whether proteins have folded correctly. The method works by fusing the gene for GFP to that of the protein being studied. It relies on the fact that the GFP will not fold correctly if the protein it is bound to folds incorrectly. Therefore, misfolded proteins will be bound to misfolded, nonfluorescing GFP, while correctly folded proteins will be signaled by fluorescent GFP. One of the advantages of the method is that it can tell you whether a protein is folded correctly, even if you don't know the function of the protein. You just need to know the location of its gene. According to scientist Geoffrey Waldo, "This assay will be particularly useful to medical researchers doing drug development and also to scientists working in the emerging field of proteomics—the study of the structures and functions of all the proteins encoded by the genome." The method is one of many developed at Los Alamos to meet the challenges facing proteomics and understanding the diseases of the brain.[20]

Guy Caldwell was studying early onset dystonia, a severe hereditary movement disorder, when he stumbled on what he hopes might be a "molecular clog remover" that could remove misfolded proteins from brain cells. The mutated gene TORA1 has been linked to early onset dystonia. Caldwell expressed the protein coded for by TORA1 in *C. elegans* to see what the function of the protein called torsin A was. Some proteins in *C. elegans* were labeled with GFP. The worm was then induced to create aggregates of improperly folded proteins with GFP tags. When torsin A was expressed by the worm, the fluorescent clusters of misfolded proteins disappeared. Torsin A was thus acting like a "Roto-rooter," removing misfolded proteins. However, in *C. elegans* with the TORA1 mutation that is known to occur in early onset dystonia, the mutated torsin A does not function properly, and the misfolded proteins do not get vacuumed up.[21] Besides having found the cause for early onset dystonia, Caldwell hopes that he might have stumbled across a protein that can remove aggregates of the destructive alpha-synuclein that are found in the brains of Parkinson's patients. Early results indicate that at least in *C. elegans*, it works: torsin A chews up the malformed alpha-synuclein clusters.

One-third of the American population is seriously overweight and termed "clinically obese." Being overweight is more than just a cosmetic problem—it can lead to severe health problems such as high blood pressure, diabetes, cardiovascular disease, stroke, and cancer. The human body has thirty billion to forty billion fat cells, which store variable amounts of fat and give it the ability to accommodate some extra fat.

Why are some people overweight and others not? Is it just diet, or are there other factors, such as a genetic propensity to putting on weight, that are responsible for the shapes of our bodies? There are no straightforward answers to these questions.

The number of overweight and obese Americans has continued to increase since 1960, a trend that is not slowing down. Part of the reason for this is a lack of physical activity combined with high-calorie, low-cost foods. However, obesity is no longer limited to Western industrialized countries; it is spreading to the developing world. Junk foods in the South Pacific Islands, Guatemala, and the deserts of Australia are cause for numerous expanding girths. In Rarotonga, the capital of the Cook Islands, 52 percent of men were clinically obese in 1996, compared with only 14 percent in 1966. Presumably these populations with a genetic propensity to gaining weight may have been held in check by diet and physical activity until now.

Determining whether someone is overweight or obese is not easy.

Doctors often use the body-mass index (BMI), which is calculated by dividing a person's weight in kilograms by his height in meters squared. According to the World Health Organization, a person with a BMI of 25 to 29 is overweight, while someone with a BMI over 30 is obese. Sounds easy enough to determine, doesn't it. Unfortunately, there are some problems. The BMI classification is largely based on mortality statistics from European and American populations. Recent studies have shown that Asians have different body types and have a particularly high risk of type 2 diabetes, cardiovascular disease, and mortality from other causes at relatively low BMIs, probably because they have a higher proportion of body fat around their waists.

One of the main culprits in our weight increase are soft drinks and sodas. The soda industry produces about 110 billion liters of soft drinks a year, and we drink it all. In fact, the average person in North America drinks about sixteen ounces a day—that's about twice as much as in 1975, and that's too much. Teenage boys drink the most soda, averaging just less than three cans a day. Each can contains about seven teaspoons of sugar, which account for the 120 calories found in most cans of soda. This is a lot of calories. What's worse is that the human body differentiates between calories obtained from soft drinks and calories from other sources. During one study, subjects had to eat jellybeans amounting to 450 calories a day for four weeks, followed by four weeks in which they consumed daily 450 calories' worth of soft drinks. The researchers in charge of the study found that the subjects ate 450 calories less of other food during their four weeks of jellybean consumption but that the soft drink intake did not affect the amount of other foods they ate.[22] Another study showed that people drinking sugar-sweetened sodas with their meals actually ate more than those drinking water or other calorie-free beverages. If the sugar consumed while drinking soda is not used fairly quickly, the chemical systems in the body start converting the sugars to other molecules that are more suited to long-term energy storage, namely, fats—and that's a problem.

Fat is important for storing energy and insulating the body against shock and temperature fluctuations, among other functions. We need fat in our food and want it because it absorbs "tasty" molecules, making the food we eat taste good. Over the last ten years, a number of hormones have been found that control the amount of food we eat. They are responsible for making us feel hungry when we need more food and for the feeling of fullness when we have eaten enough. They control our body weight. The most famous of these hormones is leptin, which tells the brain when fat stores run low and triggers appetite. The concentration of leptin in the blood is directly correlated to fat stores in lean and obese people. Lean people have less leptin in their blood, which should make them hungry. Overweight people have more leptin; as a consequence, they should feel full and not hungry. It has often been suggested that one of the genetic causes of obesity is leptin resistance; perhaps leptin does not induce the feeling of fullness in some overweight leptin-resistant people.

In March 2003, Miriam Friedman-Einat of the Institute of Animal Science at the Volcani Center in Israel published the first study that attempted to see whether the intrinsic activity of leptin could be different in overweight and lean people.[23] The leptin activity was measured by tagging the gene of a leptin-receptor protein with luciferase. If the leptin activity is high, then one should expect it to result in the expression of many leptin-receptor proteins. If, on the other hand, the leptin is hardly active, it won't be binding to many receptors, and there would be very little luciferase bioluminescence from the tagged leptin-receptor proteins. Serum levels of twenty obese and twenty nonobese individuals with a range of leptin concentration levels were tested. No difference in leptin activity was observed between the obese and nonobese patients.[24] Though this particular study didn't crack this enigma, these or similar methods likely will.

Melanocortin is another hormone that is actively involved in regulating our feelings of hunger. Many pharmaceutical companies are very interested in melanocortin because it can override the action of other hunger regulators. As noted we should expect a lean mouse to have low levels of leptin and therefore feel hungry and consequently eat more. This is true, except in melanocortin-free mice that will eat the same as normal mice despite low levels of leptin. Recently, mutations to the melanocortin receptor have been discovered that cause obesity in up to 5 percent of the human population. It has been found that melanocortin stimulates certain hypothalamic neurons by binding to the melanocortin-4-receptor, which inhibits feeding behavior. The development of drugs based on the melanocortin-4-receptor has been hampered by some of its side effects, the most interesting of which are persistent erections found in men taking drugs designed to inhibit the melanocortin-4-receptor. It probably won't surprise you that these compounds are now in development as alternatives to Viagra.

In order to examine the role of melanocortin signaling in regulating feeding behavior, Hongyan Liu of the Molecular Genetics Department at Rockefeller University has generated a line of transgenic mice that express GFP under control of a melanocortin-4-receptor promoter. To date they have only been used to establish which other neurotransmitters and small neuro-proteins hang out with melanocortin-4 receptors.[25] But someday the researchers hope to use this information to prevent obesity in some patients.

This chapter has just presented a smattering of the many different ways in which glowing gene technology has been used in the study of diseases. It is unlikely that glowing genes will be used to combat human diseases in the near future, but we will certainly benefit from the knowledge gained by using glowing genes in the examination of the causes and progression of diseases. Soon there will be many pharmaceuticals that were developed, and whose efficiency was tested, using glowing-gene technology.

CHAPTER 15

DEFENSE, SECURITY, AND BIOTERRORISM

I t has been suggested that fireflies were responsible for changing the course of Cuban history, as noted earlier. For in 1634 when British forces were sailing past the Cuban shores, they saw some flickering lights. Thinking that the Spanish fleet had beaten them there, they sailed on past Cuba. Historical records have shown that there were no Spanish forces in the area at the time; perhaps some fireflies chased off the British.[1] I doubt that any modern military force is planning to use fireflies or any other bioluminescent organisms to scare off potential invaders, but they have other plans for glowing genes.

In the second war, the Japanese used the bioluminescence of crushed mollusks to move through the jungles at night without losing each other or being seen by their opponents. The crushed snails were smeared on the backs of the foot soldiers so that they could follow each other in single file without giving off large amounts of light. Now there's an application that could be modified slightly in order for it to be used in a modern-day James Bond film. Imagine having a bioluminescent powder that emits light in the infrared region so that it is invisible to the naked eye but can be seen with special glasses. James could follow a bad guy who has touched the powder with no problem, as he would stand out like Rudolph with his red nose. In fact, that is exactly what happened to the German submarine U-34. It was sunk because its underwater movement disturbed dinoflagellates, such as those shown in figure 1 in the photo insert, that emitted some light in response to the movement, thereby giving away the position of the submarine. Even today satellites, ships, and aircraft are used to detect the telltale luminescent trails of submarines, and many tax dollars are being spent to try to hide the glow that can be observed in the wake of submerged submarines.

Besides giving away the position of submarines, bioluminescent light emission can also be used to find buried landmines. *Pseudomonas putida* are bacteria that use trinitrotoluene (TNT) as a food source. The GFP gene has been attached to the gene for the *P. putida* protein that is responsible for digesting the explosive.[2] The transgenic bacteria can then be used as sensors for explosives containing TNT.

In the last twenty-five years, as many as 200 million landmines have been produced, mainly by China, Italy, the former Soviet Union, and the United States. They remain a threat in dozens of countries. In Cambodia alone, there are thirty-five thousand landmine amputees—the "lucky" ones who survived. Even if landmines are never buried again, the existing ones will continue to kill and maim people for many years to come. Finding landmines is expensive and dangerous work. There are more than 350 different types of antipersonnel mines. How can they safely be found? Many are made of wood and plastic and can't be found using metal detectors. The potential for using transgenic bacteria as landmine detectors was tested at the National Explosives Waste Technology and Evaluation Center in South Carolina, where five deactivated test landmines were buried in a quarter-acre field. The bacteria were sprayed over the field in the hope that they would detect and then digest the TNT that had diffused from the five landmines. A UV light, suspended fifteen feet above the minefield from a cherry picker, was shone over the ground to find any fluorescence due to the GFP-tagged digestion enzyme in the bacteria. The experiment was only a partial success. Bacterial fluorescence was observed at all five antipersonnel mines. However, two false positives were detected, showing landmines where there were none.

A plant-based TNT detection system has also been tested, but it is a time-consuming and expensive method. Plant seeds are evenly spread over the minefields using a helicopter. The plant roots absorb the TNT, and GFP is expressed in the leaves if TNT is present. Once the mines have been deactivated, the transgenic TNT-detecting plants can be removed simply by uprooting all of them, thereby curtailing the spread of transgenic organisms.[3] Since plants are a lot easier to spot than bacteria, this is a major advantage of the plant-based TNT-detection systems.

As we have discussed earlier, glowing genes have been inserted into tadpoles so that they would glow in the presence of toxic heavy metals and into bacteria to detect arsenic and TNT. Exactly the same technology could be used to detect chemical warfare agents. In fact, this has already happened; organophosphorous hydrolase, an enzyme that is capable of degrading a variety of pesticides and nerve agents, such as soman, sarin, and VX, has already been tagged with GFP. The enzyme was labeled with GFP in order to have a quick and easy method to see whether it were still actively degrading the pesticide. No fluorescence—no activity.

The military and geopolitical landscape of the world has dramatically changed in the last few years and with it has come a whole new set of defense priorities. One of these is to find a quick, reliable, and inexpensive test for potential bioterrorism agents. In July 2003, Todd Rider and a number of coworkers at MIT reported a sensor that could rapidly screen samples for the low-level presence of a variety of pathogens, such as those responsible for anthrax, smallpox, SARs, Legionnaire's disease, and the plague.[4] The sensors are based on aequorin light emission and are named CANARYs (cellular analysis and notification of antigen risks and yields). They use genetically modified B lymphocytes—white blood cells that belong to the adaptive immune system and have evolved to identify pathogens very efficiently. When a pathogen binds to an antigen on the white blood cell's surface, it releases a calcium signal, which causes the cells to go on the "warpath" against the invaders. Rider, a thirty-five-year-old Arkansas native, has engineered the white blood cells so that they also express aequorin as well as membrane-bound antibodies specific for the pathogens of interest. Using CANARYs with antibodies that are sensitive to anthrax spores, Rider has been able to detect very low levels of anthrax spores in nasal passages. He hopes that the sensors will be used as first-response detectors in a variety of situations, such as testing mucus supplied by a mysteriously ill patient and analyzing an unknown white powder sent in the mail. The US Defense Advanced Research Projects Agency (DARPA) funds the research, which started in 1997. Although the blue glow emitted by aequorin is too low in intensity to be visible to the naked eye and the aequorin calcium reaction does not occur fast enough without the use of a small centrifuge, Rider thinks the technology is ready for field use. In 2004, the current prototype, which includes a centrifuge, luminometer, and computer, can be used as a field-testing kit. It is about the size of a suitcase, and the hardware costs less than $10,000.

Recently anthrax has been the object of a great deal of media attention because of its use as a bioterrorism weapon. In October 2001, eleven people inhaled anthrax spores through contact with contaminated mail; of these five people died. *Bacillus anthracis* causes anthrax in animals and humans.

Raymond Schuch and Vincent Fischetti have used a different approach to detect anthrax spores. In research also funded by DARPA they discovered an enzyme, PlyG lysine, that kills *Bacillus anthracis*. It does this by breaking open the cell wall of the anthrax bacteria. The enzyme is found in a bacteria-eating virus that specifically infects *B. anthracis*. In the August 2002 issue of *Nature*, Schuch and Fischetti reported that a single dose of PlyG lysin saved more than 80 percent of mice infected with anthrax from dying and that it could be used as a rapid test for anthrax.[5] The test is based on the fact that *B. anthracis* contains a fair amount of ATP. When a suspicious sample needs to be tested, PlyG lysin can be added to the white powder. If the powder contains anthrax spores, their cell walls will be broken by the PlyG lysin, releasing ATP, which can be detected with a luciferase assay. Since the enzyme is very specific, it will break down only anthrax cell walls, and the ATP/luciferase light emission will only occur in the presence of *B. anthracis*.

The US Army Medical Research Institute of Infectious Diseases has developed mutant *Bacillus anthracis* that express GFP. The presence, germination, and replication of these anthrax bacteria could be observed by their fluorescence in living cells. The bacteria are currently being used to help identify novel anthrax virulence factors and assess host resistance, vaccine candidates, and antianthrax therapeutics.[6]

"Gulf War syndrome" has been in the news a lot during the last ten years. Many soldiers who returned from the Gulf War have had an unexplained increase in headache and memory loss, fatigue, sleep disorders, and intestinal and respiratory ailments. These symptoms have all been blamed on exposure to some unknown chemicals during the Gulf War. There is, however, some doubt about whether the rate of symptoms among Gulf War veterans is significantly higher than among military service members who did not go to the Gulf War.[7] On the other hand, there is no doubt that soldiers in Vietnam were exposed to a very toxic chemical and that many of them are suffering the consequences. The chemical is dioxin, a component of Agent Orange. First, I will give you some background information about dioxin and Agent Orange, and then I will describe a luciferase-based test that is used as a low-cost method for the screening of dioxin exposure in studies of Vietnam veterans.

Scientists have known for a long time that plants have hormones that control their growth. The first hormone to be isolated and characterized was auxin in 1928. Scientists were very interested in the properties of auxin and tried to synthesize analogs with the hope that these could make plants grow faster and bigger than they normally do. Two of the auxin analogs that were made in the lab were 2,4-dichlorophenoxyacetic acid (2,4-D) and 2,4,5-trichlorophenoxyacetic acid (2,4,5-T). Both were very effective; they in fact induced such rapid growth in broad-leaf plants that they grew themselves to death. Pretty soon 2,4-D and 2,4,5-T weren't being used as growth stimulants; instead, they were being used as pesticides on lawns and golf courses.[8]

During the Vietnam War, the Vietcong were using the protection of the jungle foliage to mount hit-and-run attacks. The US army scientists came up with a special mixture of 2,4-D and 2,4,5-T that was particularly efficient at killing the trees and plants that the Vietcong were using as cover. The pesticide was shipped over to Vietnam in large, brightly colored barrels, which were responsible for the pesticide's common name "Agent Orange." Neither 2,4-D nor 2,4,5-T were known to cause any health problems, and so in an undertaking called Operation Ranch Hand, the military sprayed forty-two million liters over the Vietnamese jungle, the Vietcong, and American soldiers. Halfway through the war in 1969, scientists back in America found out that a contaminant produced in the manufacture of 2,4,5-T caused birth defects in humans. Knowing that many soldiers had been exposed to Agent Orange and the contaminant known as dioxin, more research was initiated to investigate the effects of dioxin. The initial results were alarming. Tests on guinea pigs showed that dioxin was the most toxic man-made chemical tested. According to Joe Schwarcz, a chemist and one of my favorite science writers, "A millionth of a gram could kill a guinea pig. Arsenic and cyanide seemed like candy by comparison."[9] As you can imagine, these findings created quite some interest, concern, and, of course, more research. Soon dioxin studies were being conducted all over, and the results were rather confusing. Guinea pigs were extremely sensitive to dioxin. Mice, rabbits, and monkeys were about two hundred times less sensitive, and hamsters, which to me look very much like guinea pigs, were two thousand times less sensitive. How susceptible to dioxin are humans? We don't know. All we do know is that we aren't as sensitive to dioxin as guinea pigs and that long-term exposure to dioxin causes cancer in humans as well as all other species of animals.

Dioxin is an extremely potent carcinogen. What makes matters worse is that dioxin is fat soluble, which means that once it enters our bodies, it is there to stay, and it accumulates over time. Every human being on Earth has some dioxin in his body fat, probably not enough to affect him, but certainly enough to measure. It is found in the fat of fish, cattle, milk, pigs, and chickens. Since 95 percent of the dioxins found in humans come from animal fats, the risk of dioxin can be reduced by eating fewer animal fats or by becoming a vegetarian. The dioxins don't all come from Agent Orange or pesticides; in fact, that is a minor source of dioxin in people who were not soldiers in the Vietnam War. Most dioxin comes from burning garbage that contains chlorinated materials such as plastics and from the paper-bleaching process. Fortunately, the levels of dioxin released into the environment are steadily dropping. This is because modern incinerators burn their garbage at much higher temperatures than the old ones and produce much less dioxin and because the use of chlorine in the bleaching of paper is being more closely regulated.

Most current methods for the detection of polycyclic and halogenated aromatic hydrocarbons, such as dioxin, are rather expensive and slow. Researchers at the University of California, Davis, have come up with an alternate, cheaper, and faster test that is based on firefly luciferase expression. It is called the CALUX (Chemical-Activated LUciferase gene eXpression cell bioassay system) assay.[10] Drs. George C. Clark and Michael S. Denison, who did most of the research, started a biotech company, Xenobiotic Detection Systems, based on their assay for dioxin and dioxinlike molecules. In 1998 they were awarded a patent for CALUX. The Belgium government's Scientific Institute of Public Health, Hiyoshi Corporation of Japan, and the Federal Drug Administration have all signed licensing agreements with Xenobiotic Detection Systems to use the CALUX system to test for dioxinlike molecules in foods, animal feeds, and tissue. The Belgium government is particularly interested in the CALUX system because Belgium had a major dioxin scare in 2001, when chickens began dying due to dioxins in their food. Some shady chicken-feed manufacture had purportedly mixed transformer fluid containing dioxin with the animal fat in their chicken feed and had thereby contaminated all their feed.[11]

There are probably many other applications of glowing-gene technology in the defense, security, and bioterrorism area that I know nothing about and that have not been publicized because they would be much more effective if terrorists did not know about them.

CHAPTER 16

LAST LIGHT

Many years ago, Osamu Shimomura and Bill McElroy started doing some basic research. Their goal wasn't to find a new way of seeing how and when proteins are made inside living organisms. They were just trying to find out how the crystal jellyfish (*Aequorea victoria*) and the firefly managed to emit light. When Bill McElroy spent large portions of his life and significant amounts of grant money to determine how fireflies generated light, he didn't foresee any direct uses for this knowledge. Osamu Shimomura caught millions of jellyfish and painstakingly isolated the materials that were responsible for its bioluminescence. He was doing basic research—he wanted to know how *Aequorea victoria* produced green light when agitated. It is likely that he never considered the possibility that the proteins responsible for the green light would be used to track the growth of blood vessels to cancerous cells, even when he was ruminating over his laboratory problems in a rowboat.

One of the most common questions scientists are often asked about their research is "Why are you doing this? What is the practical application of your research?" The answer that it is basic research and that we would like to increase our knowledge of science often is not very satisfying to most nonscientists. It is a bit like climbing mountains just because they are there. To the mountaineer, that is a perfectly good and logical reason; to me, it seems to be a little odd and not much of a reason to risk one's life.

Basic research produces ideas about scientific phenomena—ideas that, if they are good, can be published in the scientific literature. We have no way of knowing which ideas will end up overcoming all hurdles and trigger a scientific revolution. There is no way to determine which basic research will eventually lead to a major breakthrough that will revolutionize an area of science. The only way to move science forward is to fund and do research in many areas so that one of them can change how science is done and improve our everyday lives.

Once a baby has been born, it is subjected to many people, stimuli, and influences. It soon learns to crawl, walk, and even talk. It goes to school and learns how to read and write, to play sports, and to drive its parents crazy. The glowing-gene revolution has gone through these stages, too. First, GFP and luciferase were used only by a small number of researchers, who sorted out many of the problems. They cloned luciferase and made brighter forms of GFP, red luminescent luciferase, and yellow fluorescent protein. Now glowing genes are used at every major research university in the world; they can be purchased in kit form, and all students graduating in the biological sciences will have learned about them or even used them in their teaching laboratories. At Pomona College, for example, biologist Clarissa Cheney has a stock of GFP fruit fly larvae that express GFP in the salivary gland. It has strong fluorescence and is easy to see. Seeing the glowing larvae in their first week of labs gets students excited about the course. If there are students interested in dissecting salivary glands, these fruit flies are used. The students can then check whether or not they have cut out the salivary glands by seeing if the tissue they dissected fluoresces. Cheney also has another stock of fruit flies that express GFP in their nervous system. She said, "They are visually captivating. You can see the larvae, and even adult flies, moving around, while seeing exactly where the brain and major nerves are."[1]

Glowing genes have revolutionized biotechnology. They are used on a daily basis and allow us to see a whole new world that we would not have been able to examine without them. However, glowing-gene technology is, in effect, in its teenage years. We don't know for certain what it will do in the next few years. Will it go to college and make its mark on the rest of the world, or will it continue being the apple of its families' eye without ever breaking into the big wide world? Only time will tell whether GFP and luciferase will make the transition from a revolutionary tool in biotechnology to a technique that is capable of revolutionizing all of science and maybe even life. How long will it take before we can tell whether the glowing-gene revolution transcends even a revolution in biotechnology? I don't know. Scientific revolutions generally take a long time, but the pace of science is rapidly increasing. Let's just compare glowing genes with the advent of the microscope.

More than 240 years ago, Anton van Leeuwenhoek was the first to study bacteria using a microscope. His microscopes had a magnifying power of up to 270 times larger than actual size. I think even he would be surprised to learn that today nearly every school has a microscope, that children are given microscopes as presents, that conventional microscopes with magnifications of two thousand are commonly available, and that scanning-tunneling microscopes with magnifications of 100 million are able to "see" individual atoms.

When Leeuwenhoek first described his observations of bacteria and sperm, it took a while before people first heard about it. Today, with international conferences and online publishing, information flows freely and fast. Green fluorescent protein was first used as a marker protein ten years ago, and now it is extremely rare to find a copy of *Science*, *Nature*, or the *Proceedings of the National Academy of Sciences* that does not contain an article that uses GFP or one of its analogs. What will we be able to see using bioluminescent gene technology fifty years from now? Will our grandchildren have chemistry/biotech kits to virally infect different species of pets with viruses that encode for brightly colored bioluminescent proteins?

Thanks to the development of the microscope, scientists in the nineteenth century were able to see what is inside a cell. They were able to provide descriptions of the structures found inside the cell. But they could not determine the chemical nature of the structures or the reactions that occur inside them. In the twentieth century, genetics and biochemistry took over for the good old microscope, and scientists began to understand the underlying chemistry of many of the important processes in the cell. One of the largest and most significant projects of the century was the determination of the human genome. Sequencing the complete human genome was an amazing piece of work, but it was only the start because the genomic sequence lacks spatial and temporal information. According to Roger Tsien, it is as dynamic and informative as a census list or telephone directory.[2] He thinks that the challenge for the twenty-first century is to understand how the proteins described in the human genome work together to make living cells and organisms and to use that information to improve health and well-being.[3] GFP, firefly luciferase, and the other bioluminescent proteins that have yet to be found will play vital roles in this challenge for the twenty-first-century scientist. It is our hope that they will provide the temporal and spatial information we need to take us from a telephone book to a novel, a play, or even a film. By using glowing genes, we hope that scientists will be able to map out all the complex interactions that occur in the life of a protein. We will then know more than just the gene's address on the chromosome; we will know how, when, and why the protein coded by the gene is made, where it goes, and what other proteins interact with it.

Lynn Cooley, a professor at the Yale University School of Medicine, has taken the first steps toward this goal. She has established Flytrap, a Web-based database that contains information generated by tagging every gene in the fruit fly with the GFP gene.[4]

Glowing genes have revolutionized biotechnology. Now that you have read about these innovations and techniques, you will appreciate all the work that went into getting us to this point. And just think of all future applications we have yet to see.

NOTES

INTRODUCTION

1. Bill Bryson, *A Short History of Nearly Everything* (New York: Broadway Books, 2003), pp. 377–80.

2. A. J. Trewavas and C. J. Leaver, "Is Opposition to GM Crops Science or Politics? An Investigation into the Arguments That GM Crops Pose a Particular Threat to the Environment," *European Molecular Biology Organization Reports* 2, no. 6 (2001): 455–59.

3. Joe Schwarcz, *That's the Way the Cookie Crumbles* (Toronto: ECW Press, 2002).

4. R. L. Rawls, "Tackling Arsenic in Bangladesh," *Chemical and Engineering News* 80, no. 42 (2002): 42–45.

5. J. Stocker, D. Balluch, M. Gsell, H. Harms, J. Feliciano, S. Daunert, K. A. Malik, and J. R. Van der Meer, "Development of a Set of Simple Bacterial Biosensors for Quantitative and Rapid Measurements of Arsenite and Arsenate in Potable Water," *Environmental Science and Technology* 37, no. 20 (2003): 4743–50.

1. LIVING LIGHT

1. Lars O. Bjorn, *Photobiology: The Science of Light and Life* (Dordrecht: Kluwer, 2002), pp. 389–99.

2. *Monterey Bay Aquarium*, http://www.mbayaq.org/ [accessed August 27, 2003].

3. Ibid.

4. M. A. Cody, "The Physiology of Bioluminescence in Marine Animals," http://www.bio.davidson.edu/courses/anphys/1999/Cody/Cody.htm [accessed August 27, 2003].

5. Bill Bryson, *A Short History of Nearly Everything* (New York: Broadway Books, 2003), pp. 274–79.

6. Ibid.

7. Ibid.

8. Catherine L. Hines, *The Official William Beebe Web Site*, http://members.aol.com/chines6930/mw1/bio.htm [accessed August 28, 2003].

9. Ibid.

10. Bryson, *Short History of Nearly Everything*, pp. 274–79.

11. Jennifer Uscher, "Deep-Sea Machine," *PBS*, http://www.pbs.org/wgbh/nova/abyss/frontier/deepsea.html [accessed August 28, 2003].

12. Bryson, *Short History of Nearly Everything*, pp. 274–79.

13. S. J. Bourlat, C. Nielsen, A. E. Lockyer, D. T. J. Littlewood, and M. J. Telford, "Xenoturbella Is a Deuterostome That Eats Molluscs," *Nature* 424, no. 6951 (2003): 925–28.

14. Lazzaro Spallanzani, *Viaggi alle due Sicilie* (Pavia, Italy: Stamperia di B. Comini, 1793).

15. Bernhard Grzimek, *Grzimek's Animal Life Encyclopedia*, vol. 2, *Insects* (Zurich: Van Nostrand Reinhold, 1972).

16. Ibid.

17. Spallanzani, *Viaggi alle due Sicilie*.

18. F. V. Vencl and A. D. Carlson, "Proximate Mechanisms of Sexual Selection in the Firefly Photinus Pyralis," *Journal of Insect Behavior* 11 (1997): 191–207; C. K. Cratsley, J. A. Rooney, and S. M. Lewis, "Limits to Nuptial Gift Production by Male Fireflies, *Photinus Ignitus*," *Journal of Insect Behavior* 16, no. 3 (2003): 361–70.

19. Cratsley, Rooney, and Lewis, "Limits to Nuptial Gift Production by Male Fireflies, *Photinus Ignitus*," pp. 361–70.

20. W. W. Ward, "Fluorescent Proteins: Who's Got 'Em and Why?" in *Bioluminescence and Chemiluminescence*, ed. P. E. Stanley and L. J. Kricka (Singapore: World Scientific, 2002), p. 123.

21. Grzimek, *Grzimek's Animal Life Encyclopedia*, vol. 2; T. Eisner, M. A. Goetz, D. E. Hill, S. R. Smedley, and J. Meinwald, "Firefly 'Femmes Fatales' Acquire Defensive Steroids (Lucibufagins) from Their Firefly Prey," *Proceedings of the National Academy of Sciences of the United States of America* 94, no. 18 (1997): 9723–28

22. J. E. Lloyd, "Evolution of a Firefly Flash Code," *Florida Entomologist* 67 (1984): 368–76.

23. Scott Camazine, Jean-Louis Deneubourg, Nigel R. Franks, James Sneyd, Guy Theraulaz, and Eric Bonabeau, *Self-Organization in Biological Systems* (Princeton, NJ: Princeton University Press, 2001).

24. Matthew Bennett, Michael F. Schatz, Heidi Rockwood, and Kurt Wiesenfeld, "Huygens's Clocks," *Proceedings: Mathematical, Physical and Engineering Sciences* 458, no. 2019 (March 2002): 563–79.

25. Elizabeth Tayntor Gowell, *Sea Jellies: Rainbows in the Sea* (New York: Franklin Watts, 1993).

26. *Monterey Bay Aquarium.*

27. Gowell, *Sea Jellies: Rainbows in the Sea.*

2. FROM PLINY'S WALKING STICK TO BURNING ANGELS

1. Polly Shulman, "Return of the Blob (Jellyfish as Food)," *Discover* 16, no. 7 (1995): 42–46.

2. E. N. Harvey, *A History of Luminescence from the Earliest Times until 1900* (Philadelphia: American Philosophical Society, 1957).

3. Ibid.

4. Ibid.; J. Legge, "The Odes of Pin," in *The Chinese Classics*, vol. 4 (Oxford, UK: Clarendon Press, 1893), p. 237, III, verse 2.

5. Fortunius Licetus, *Litheosphorus Sive De Lapide Bononiensi* (Bologna, Italy: University of Bologna, 1640).

6. Aldo Roda, *Bioluminescence and Chemiluminescence: Perspectives for the 21st Century, Proceedings of the 10th International Symposium on Bioluminescence and Chemiluminescence Held at Bologna, Italy*, ed. Roda, M. Pazzagli, L. J. Kricka, and P. E. Stanley (Chichester, UK: Wiley, 1998).

7. Harvey, *History of Luminescence from the Earliest Times until 1900.*

8. Benjamin Franklin, *Experiments and Observations on Electricity, Made at Philadelphia in America*, 4th ed. (London: Printed for David Henry, and sold

by Francis Newbery, 1769); Harvey, *History of Luminescence from the Earliest Times until 1900*.

9. Ibid.

10. J. Rohr, M. I. Latz, S. Fallon, J. C. Nauen, and E. Hendricks, "Experimental Approaches towards Interpreting Dolphin-Stimulated Bioluminescence," *Journal of Experimental Biology* 201, no. 9 (1998): 1447–60.

11. Samuel Purchase, *Purchas His Pilgrimes* (London: Imprinted for H. Fetherston, 1625).

12. Francis Bacon, *Sylva Sylvarum* (London: Printed by J.H. for William Lee, 1627).

13. Francine Jacobs, *Nature's Light: The Story of Bioluminscence* (New York: Morrow, 1974).

14. Bill Bryson, *A Short History of Nearly Everything* (New York: Broadway Books, 2003).

15. Bruce Jones, *Life and Letters of Faraday* (London: Longmans, Green, 1870).

16. Bryson, *Short History of Nearly Everything*.

17. Athanasius Kircher, *Ars Magna Lucis Et Umbrae*, ed. and trans. R. A. Applegate (Amstelodami, apud Joannem Janssonium à Waesberge & hæredes Elizæi Wayerstraet, 1671).

18. Harvey, *History of Luminescence from the Earliest Times until 1900*; A. von Humboldt and A. Bonpland, *Personal Narrative of Travels to the Equinoctial Regions of America, during the Years 1799–1804* (London: G. Bell & Sons, 1881).

19. Lazzaro Spallanzani, *Viaggi alle due Sicilie* (Pavia, Italy: Stamperia di B. Comini, 1793).

20. Harvey, *History of Luminescence from the Earliest Times until 1900*; Spallanzani, *Viaggi alle due Sicilie*.

21. Alphonse Leroy, *Philosophical Magazine* 2 (1798): 290–93.

3. USING FIREFLIES TO LOOK FOR LIFE ON MARS?

1. "Bioluminescence Expert William McElroy Dies at 82," *Johns Hopkins University Gazette* 28, no. 24 (March 1, 1999).

2. "William D. McElroy, Scientist, Dies at 82," *Washington Post*, February 20, 1999, p. B6.

3. "Obituary for former UC San Diego Chancellor and National Science Foundation Director William D. McElroy," *University of California, San Diego,*

http://adminrecords.ucsd.edu/Notices/1999/1999-02-18-2.html [accessed September 1, 2003].

4. Frederick N. Rasmussen, "He Shed Light on Glowing Insects," *Baltimore Sun*, June 12, 1999, p. 8E.

5. Bernhard Grzimek, *Grzimek's Animal Life Encyclopedia*, vol. 2, *Insects* (Zurich: Van Nostrand Reinhold, 1972).

6. Rasmussen, "He Shed Light on Glowing Insects," p. 8E.

7. Ibid.

8. J. W. Hastings, "Firefly Flashes and Royal Flushes: Life in a Full House," *Journal of Bioluminescence and Chemiluminescence* 4, no. 1 (1989): 29–30.

9. Ibid.

10. Ibid.

11. D. W. Ow, K. V. Wood, M. DeLuca, J. R. Dewet, D. R. Helinski, and S. H. Howell, "Transient and Stable Expression of the Firefly Luciferase Gene in Plant Cells and Transgenic Plants," *Science* 234 (November 14, 1986): 856–59.

12. Bill Bryson, *A Short History of Nearly Everything* (New York: Broadway Books, 2003), pp. 377–80.

13. K. Venkateswaran, N. Hattori, M. T. La Duc, and R. Kern, "ATP as a Biomarker of Viable Microorganisms in Clean-Room Facilities," *Journal of Microbiological Methods* 52, no. 3 (March 2003): 367–77.

14. Tom Clarke, "The Stowaways," *Nature Science Update 2001*, http://www.nature.com/nsu/010920/010920-16.html [accessed November 11, 2003].

15. Hastings, "Firefly Flashes and Royal Flushes," pp. 29–30.

16. B. A. Trimmer, J. R. Aprille, D. M. Dudzinski, C. J. Lagace, S. M. Lewis, T. Michel, S. Qazi, and R. M. Zayas, "Nitric Oxide and the Control of Firefly Flashing," *Science* 292, no. 5526 (2001): 2486–88.

17. Grzimek, *Grzimek's Animal Life Encyclopedia*, vol. 2.

4. SHIMOMURA'S "SQUEEZATE"

1. O. Shimomura, "A Short Story of Aequorin," *Biological Bulletin* 189, no. 1 (1995): 1–5.

2. O. Shimomura, F. H. Johnson, and Y. Saiga, "Extraction, Purification and Properties of Aequorin, a Bioluminescent Protein from the Luminous Hydromedusan, Aequorea," *Journal of Cellular and Comparative Physiology* 59 (June 1962): 223–29.

3. A. von Humboldt and A. Bonpland, *Personal Narrative of Travels to the Equinoctial Regions of America, during the Years 1799–1804* (London: G. Bell & Sons, 1881).

4. S. P. Colin and J. H. Costello, "Morphology, Swimming Performance and Propulsive Mode of Six Co-Occurring Hydromedusae," *Journal of Experimental Biology* 205, no. 3 (2002): 427–37.

5. H. Morise, O. Shimomura, F. H. Johnson, and J. Winant, "Intermolecular Energy Transfer in Bioluminescent Systems of Aequorea," *Biochemistry* 13 (1974): 2656–62.

6. S. Forsen and J. Koerdel, "Calcium in Biological Systems," in *Bioinorganic Chemistry*, ed. I. Bertini, H. B. Gray, S. J. Lippard, and J. S. Valentine (Mill Valley, CA: University Science Books, 1994).

7. Ibid.

8. Ibid.

9. L. C. Mattheakis and L. D. Ohler, "Seeing the Light: Calcium Imaging in Cells for Drug Discovery," *Drug Discovery Today* (2000): 15–19.

10. A. Trewavas, "Mindless Mastery," *Nature* 415, no. 6874 (2002): 841–41.

11. Ibid.

12. M. R. Knight, S. M. Smith, and A. J. Trewavas, "Wind-Induced Plant Motion Immediately Increases Cytosolic Calcium," *Proceedings of the National Academy of Sciences of the United States of America* 89, no. 11 (1992): 4967–71.

13. M. R. Knight, A. K. Campbell, S. M. Smith, and A. J. Trewavas, "Transgenic Plant Aequorin Reports the Effects of Touch and Cold-Shock and Elicitors on Cytoplasmic Calcium," *Nature* 352, no. 6335 (1991): 524–26.

14. D. Bradley, "Shining, Unhappy Plants," *American Scientist* 84, no. 1 (1996): 25.

15. A. Falciatore, M. R. d'Alcala, P. Croot, and C. Bowler, "Perception of Environmental Signal by a Marine Diatom," *Science* 288, no. 5475 (2000): 2363–66.

16. N. T. Wood, A. Haley, M. Viry-Moussaid, C. H. Johnson, A. H. van der Luit, and A. J. Trewavas, "The Calcium Rhythms of Different Cell Types Oscillate with Different Circadian Phases," *Plant Physiology* 125, no. 2 (2001): 787–96.

17. J. Sai and C. H. Johnson, "Different Circadian Oscillators Control Ca(2+) Fluxes and *Lhcb* Gene Expression," *Proceedings of the National Academy of Sciences of the United States of America* 96, no. 20 (1999): 11659–63.

5. WHERE IS THE GFP RECIPE? LET'S PHOTO-COPY IT

1. Douglas Prasher, private communication, October 21, 2003.

2. Bill Bryson, *The Mother Tongue: English and How It Got That Way* (New York: William Morrow, 1990), p. 230.

3. D. C. Prasher, V. K. Eckenrode, W. W. Ward, F. G. Prendergast, and M. J. Cormier, "Primary Structure of the *Aequorea Victoria* Green Fluorescent Protein," *Gene* 111 (February 15, 1992): 229–33.

4. Ibid.

6. THE BIRTH OF THE GREEN FLUORESCENT PROTEIN REVOLUTION

1. Martin Chalfie, personal communication, June 10, 2003.

2. "Sydney Brenner," http://elegans.swmed.edu/Sydney.html [accessed November 5, 2004].

3. "Press Release: The 2002 Nobel Prize in Physiology or Medicine," *Nobelprize.org*, http://www.nobel.se/medicine/laureates/2002/press.html [accessed November 12, 2003].

4. Kevin Davies, *Cracking the Genome: Inside the Race to Unlock Human DNA* (New York City: Free Press, 2001), p. 91.

5. D. C. Prasher, V. K. Eckenrode, W. W. Ward, F. G. Prendergast, and M. J. Cormier, "Primary Structure of the *Aequorea Victoria* Green Fluorescent Protein," *Gene* 111 (February 15, 1992): 229–33.

6. Gary Taubes, "An Interview with Martin Chalfie, Ph.D.," *In-cites*, http://www.in-cites.com/papers/DrMartinChalfie.html [accessed November 12, 2003].

7. W. W. Ward, "Fluorescent Proteins: Who's Got 'Em and Why?" in *Bioluminescence and Chemiluminescence*, ed. P. E. Stanley and L. J. Kricka (Singapore: World Scientific, 2002), p. 123.

8. M. V. Matz, K. A. Lukyanov, and S. A. Lukyanov, "Family of the Green Fluorescent Protein: Journey to the End of the Rainbow," *Bioessays* 24, no. 10 (2002): 953–59.

9. M. Chalfie, Y. Tu, G. Euskirchen, W. W. Ward, and D. C. Prasher, "Green Fluorescent Protein as a Marker for Gene Expression," *Science* 263 (February 11, 1994): 802–805.

10. Ibid.

11. Ibid.

12. S. X. Wang and T. Hazelrigg, "Implications for Bcd mRNA Localization from Spatial-Distribution of Exu Protein in *Drosophila* Oogenesis," *Nature* 369, no. 6479 (1994): 400–403.

13. Chalfie, personal communication, June 10, 2003.

14. Bill Bryson, *A Short History of Nearly Everything* (New York: Broadway Books, 2003).

15. Chalfie, personal communication, June 10, 2003.

16. Martin Chalfie and Steven Kain, eds., *Green Fluorescent Protein: Properties, Applications, and Protocols* (New York: Wiley-Liss, 1998).

7. THIRSTY POTATOES AND GREEN BLOOD

1. Anton van Leeuwenhoek, "Letter to Royal Society, September 17, 1683" http://neurolab.jsc.nasa.gov/leeuwen.htm [accessed November 5, 2004].

2. C. Y. Chou, L. S. Horng, and H. J. Tsai, "Uniform GFP-Expression in Transgenic Medaka (*Oryzias Latipes*) at the F0 Generation," *Transgenic Research* 10, no. 4 (2001): 303–15.

3. Z. Y. Gong, H. Y. Wan, T. L. Tay, H. Wang, M. R. Chen, and T. Yan, "Development of Transgenic Fish for Ornamental and Bioreactor by Strong Expression of Fluorescent Proteins in the Skeletal Muscle," *Biochemical and Biophysical Research Communications* 308, no. 1 (2003): 58–63.

4. Les MacPherson, "If Fish Can Glow, Why Can't Schnauzers Fly?" *Star Phoenix*, December 9, 2003, p. A3.

5. Ibid.

6. Denise Flaim, "Seeing Red over Fluorescent Fish—Some Conservationists Object to Messing with Mother Nature," *Newsday*, December 30, 2003, p. B10.

7. A. Mercuri, A. Sacchetti, L. De Benedetti, T. Schiva, and S. Alberti, "Green Fluorescent Flowers," *Plant Science* 161, no. 5 (2001): 961–68.

8. "Scottish Potato Glows When It Feels Thirsty," *The (London) Times*, December 18, 2000, p. 7.

9. Amanda Onion, "One Potato, New Potato," *ABC News.com* http://abcnews.go.com/sections/scitech/CuttingEdge/cuttingedge001222.html [accessed October 1, 2003].

10. T. G. Spiro and W. M. Stigliani, *Chemistry of the Environment* (Upper Saddle River, NJ: Prentice Hall, 1996).

11. K. Oofusa, O. Tooi, A. Kashiwagi, K. Kashiwagi, Y. Kondo, M. Obara,

and K. Yoshizato, "Metal Ion-Responsive Transgenic Xenopus Laevis as an Environmental Monitoring Animal," *Environmental Toxicology and Pharmacology* 13, no. 3 (2003): 153–59.

12. L. C. Hudson, D. Chamberlain, and C. N. Stewart Jr., "GFP-Tagged Pollen to Monitor Pollen Flow of Transgenic Plants," *Molecular Ecology Notes* 1, no. 4 (2001): 321–24.

13. I. E. Alcamo, *DNA Technology: The Awesome Skill* (Dubuque, IA: William C. Brown, 1996).

14. Hudson, Chamberlain, and Stewart, "GFP-Tagged Pollen to Monitor Pollen Flow of Transgenic Plants," pp. 321–24.

15. P. Sengupta, J. H. Chou, and C. I. Bargmann, "Odr-10 Encodes a Seven Transmembrane Domain Olfactory Receptor Required for Responses to the Odorant Diacetyl," *Cell* 84, no. 6 (March 1996): 899–909.

16. Ibid.

17. Q. M. Long, A. M. Meng, H. Wang, J. R. Jessen, M. J. Farrell, and S. Lin, "GATA-1 Expression Pattern Can Be Recapitulated in Living Transgenic Zebrafish Using GFP Reporter Gene," *Development* 124, no. 20 (1997): 4105–11.

18. M. Yang, E. Baranov, A. R. Moossa, S. Penman, and R. M. Hoffman, "Visualizing Gene Expression by Whole-Body Fluorescence Imaging," *Proceedings of the National Academy of Sciences of the United States of America* 97, no. 22 (2000): 12278–82; M. Yang, E. Baranov, P. Jiang, F. X. Sun, X. M. Li, L. N. Li, S. Hasegawa, M. Bouvet, M. Al-Tuwaijri, T. Chishima, H. Shimada, A. R. Moossa, S. Penman, and R. M. Hoffman, "Whole-Body Optical Imaging of Green Fluorescent Protein-Expressing Tumors and Metastases," *Proceedings of the National Academy of Sciences of the United States of America* 97, no. 3 (2000): 1206–11.

19. N. Saito, M. Zhao, L. N. Li, E. Baranov, M. Yang, Y. Ohta, K. Katsuoka, S. Penman, and R. M. Hoffman, "High Efficiency Genetic Modification of Hair Follicles and Growing Hair Shafts," *Proceedings of the National Academy of Sciences of the United States of America* 99, no. 20 (2002): 13120–24.

20. A. Tomicka, J. R. Chen, S. Barbut, and M. W. Griffiths, "Survival of Bioluminescent Escherichia Coli O157:H7 in a Model System Representing Fermented Sausage Production," *Journal of Food Protection* 60, no. 12 (1997): 1487–92.

21. J. Reunanen and P. E. J. Saris, "Bioassay for Nisin in Sausage: A Shelf Life Study of Nisin in Cooked Sausage," *Meat Science* 66, no. 3 (2004): 515–18.

22. T. F. de Koning-Ward, C. J. Janse, and A. P. Waters, "The Development of Genetic Tools for Dissecting the Biology of Malaria Parasites," *Annual Review of Microbiology* 54 (2000): 157–85.

23. C. S. C. Price, K. A. Dyer, and J. A. Coyne, "Sperm Competition between *Drosophila* Males Involves Both Displacement and Incapacitation," *Nature* 400, no. 6743 (1999): 449–52.

8. ALBA, THE FLUORESCENT RABBIT

1. Christopher Dickey, "I Love My Glow Bunny: Genetically Modified Objet D'art? Crime against Nature? Transgenic Protein Machine? The Inside Story of How a Reengineered Rabbit Named Alba Became the Center of an Intercontinental Tug-of-War," *Wired* (April 2001).

2. Ibid.

3. Robert Silverberg, "Reflections: The Case of the Phosphorescent Rabbit," *Asimov's Science Fiction* (September 2001).

4. Brigitte Boisselier, "Glowing Controversy: Comment by Brigitte Boisselier, Ph.D., Raelian Bishop Guide," http://www.rael.org/int/portugese/news/news/article1.htm [accessed November 12, 2003].

5. Kristen Philipkoski, "RIP: Alba, the Glowing Bunny," *Wired News*, http://www.wired.com/news/medtech/0,1286,54399,00.html [accessed October 29, 2003].

6. "Transgenic Light," *Stanford University*, http://hpslab.stanford.edu:16080/projects/TransgenicLight/TransgenicLight.html [accessed October 27, 2003].

7. Ibid.

8. Holger Breithaupt, "Rhythm 'N' Biology: What Happens If Cell Biology and Techno Music Meet?" *European Molecular Biology Organization Reports* 3, no. 9 (September 2002): 813–15.

9. Ibid.

9. LIGHT IN A CAN

1. IBM (press release), "IBM Announces $100 Million Research Initiative to Build World's Fastest Supercomputer," December 6, 1999.

2. J. W. Pitera and W. Swope, "Understanding Folding and Design: Replica-Exchange Simulations of 'Trp-Cage' Fly Miniproteins," *Proceedings of the National Academy of Sciences of the United States of America* 100, no. 13 (2003): 7587–92.

3. M. M. Waldrop, "Science Behind the Screens," *HHMI Bulletin* (June 2003): 4–5.

4. C. D. Snow, N. Nguyen, V. S. Pande, and M. Gruebele, "Absolute Comparison of Simulated and Experimental Protein-Folding Dynamics," *Nature* 420, no. 6911 (2002): 102–106.

5. *Department of Physics, Cambridge University*, http://www.phy.cam.ac .uk/camphy/xraydiffraction/xraydiffraction9_1.htm [accessed March 9, 2004].

6. F. Yang, L. Moss, and G. Phillips, "The Molecular Structure of Green Fluorescent Protein," *Nature Biotechnology* 14 (1996): 1246–51; M. Ormoe, A. B. Cubitt, K. Kallio, L. A. Gross, R. Y. Tsien, and S. J. Remington, "Crystal Structure of the *Aequorea Victoria* Green Fluorescent Protein," *Science* 273 (1996): 1392–95.

7. R. M. Wachter, M. A. Elsiger, K. Kallio, G. T. Hanson, and S. J. Remington, "Structural Basis of Spectral Shifts in the Yellow Emission Variants of Green Fluorescent Protein," *Structure* 6 (1998): 1267–77.

8. L. Wang, J. M. Xie, A. A. Deniz, and P. G. Schultz, "Unnatural Amino Acid Mutagenesis of Green Fluorescent Protein," *Journal of Organic Chemistry* 68, no. 1 (2003): 174–76.

9. I. Ghosh, A. D. Hamilton, and L. Regan, "Antiparallel Leucine Zipper-Directed Protein Reassembly: Application to the Green Fluorescent Protein," *Journal of the American Chemical Society* 122, no. 23 (2000): 5658–59.

10. S. Zhang, C. Ma, and M. Chalfie, "Combinatorial Marking of Cells and Organelles with Reconstituted Fluorescent Proteins," *Cell* 119 (2004) 137–44.

11. D. Fang and T. K. Kerppola, "Ubiquitin-Mediated Fluorescence Complementation Reveals that Jun Ubiquitinated by Itch/AIP4 Is Localized to Lysosomes," *Proceedings of the National Academy of Sciences of the United States of America* 101, no. 41 (2004): 14782–87.

12. E. Conti, N. Franks, and P. Brick, "Crystal Structure of Firefly Luciferase Throws Light on a Superfamily of Andenylate-Forming Enzymes," *Structure* 4 (1996): 287.

10. RED SHEEP FROM RUSSIA

1. M. V. Matz, K. A. Lukyanov, and S. A. Lukyanov, "Family of the Green Fluorescent Protein: Journey to the End of the Rainbow," *Bioessays* 24, no. 10 (2002): 953–59.

2. Ibid.

3. M. V. Matz, A. F. Fradkov, Y. A. Labas, A. P. Savitisky, A. G. Zaraisky, M. L. Markelov, and S. A. Lukyanov. "Fluorescent Proteins from Nonbioluminescent Anthozoa Species," *Nature Biotechnology* 17 (1999): 969–73.

4. W. W. Ward, "Fluorescent Proteins: Who's Got 'Em and Why?" in *Bioluminescence and Chemiluminescence*, ed. P. E. Stanley and L. J. Kricka (Singapore: World Scientific, 2002), p. 123.

5. M. A. Wall, M. Socolich, and R. Ranganathan, "The Structural Basis for Red Fluorescence in the Tetrameric GFP Homolog DsRed," *Natural Structural Biology* 7, no. 12 (2000): 1133–38; D. Yarbrough, R. M. Wachter, K. Kallio, M. V. Matz, and S. J. Remington, "Refined Crystal Structure of DsRed, a Red Fluorescent Protein from Coral, at 2.0-Angstrom Resolution," *Proceedings of the National Academy of Sciences of the United States of America* 98, no. 2 (2001): 462–67.

6. R. E. Campbell, O. Tour, A. E. Palmer, P. A. Steinbach, G. S. Baird, D. A. Zacharias, and R. Y. Tsien, "A Monomeric Red Fluorescent Protein," *Proceedings of the National Academy of Sciences of the United States of America* 99, no. 12 (2002): 7877–82.

7. Y. A. Labas, N. G. Gurskaya, Y. G. Yanushevich, A. F. Fradkov, K. A. Lukyanov, S. A. Lukyanov, and M. V. Matz, "Diversity and Evolution of the Green Fluorescent Protein Family," *Proceedings of the National Academy of Sciences of the United States of America* 99, no. 7 (2002): 4256–61.

8. K. A. Lukyanov, A. F. Fradkov, N. G. Gurskaya, M. V. Matz, Y. A. Labas, A. P. Savitsky, M. L. Markelov, A. G. Zaraisky, X. N. Zhao, Y. Fang, W. Y. Tan, and S. A. Lukyanov, "Natural Animal Coloration Can Be Determined by a Nonfluorescent Green Fluorescent Protein Homolog," *Journal of Biological Chemistry* 275, no. 34 (2000): 25879–82.

9. A. Terskikh, A. Fradkov, G. Ermakova, A. Zaraisky, P. Tan, A. V. Kajava, X. N. Zhao, S. Lukyanov, M. Matz, S. Kim, I. Weissman, and P. Siebert, "'Fluorescent Timer': Protein That Changes Color with Time," *Science* 290, no. 5496 (2000): 1585–88.

10. M. Chicurel, "Color-Changing Protein Times Gene Activity," *Science* 290 (2000): 1478.

11. Terskikh et al., "'Fluorescent Timer': Protein That Changes Color with Time," pp. 1585–88.

12. Y. G. Yanushevich, N. G. Gurskaya, D. B. Staroverov, S. A. Lukyanov, and K. A. Lukyanov, "A Natural Fluorescent Protein That Changes Its Fluorescence Color During Maturation," *Russian Journal of Bioorganic Chemistry* 29, no. 4 (2003): 325–29.

13. Chicurel, "Color-Changing Protein Times Gene Activity," p. 1478.

11. ANDi THE GREEN MONKEY AND A YELLOW PIG

1. K. Senior, "What Next after the First Transgenic Monkey?" *Lancet* 357, no. 9254 (February 2001): 450.

2. S. B. Dunnett, "Reverse Transcription of Inserted DNA in a Monkey Gives Us ANDi," *Trends in Pharmacological Sciences* 22, no. 5 (2001): 211–15; A. W. S. Chan, K. Y. Chong, D. Takahashi, C. Martinovich, N. Duncan, L. Hewitson, C. Simerly, and G. Schatten, "Reverse Transcription of Inserted DNA in a Monkey Gives Us ANDi—Response," *Trends in Pharmacological Sciences* 22, no. 5 (2001): 214–15.

3. Bill Bryson, *A Short History of Nearly Everything* (New York: Broadway Books, 2003).

4. Ibid.

5. G. Vogel, "Transgenic Animals—Infant Monkey Carries Jellyfish Gene," *Science* 291, no. 5502 (2001): 226–26.

6. A. W. S. Chan, K. Y. Chong, C. Martinovich, C. Simerly, and G. Schatten, "Transgenic Monkeys Produced by Retroviral Gene Transfer into Mature Oocytes," *Science* 291, no. 5502 (2001): 309–12.

7. Ibid.

8. Dunnett, "Reverse Transcription of Inserted DNA in a Monkey Gives Us Andi," pp. 211–14.

9. R. S. Prather, R. J. Hawley, D. B. Carter, L. Lai, and J. L. Greenstein, "Transgenic Swine for Biomedicine and Agriculture," *Theriogenology* 59, no. 1 (2003): 115–23.

10. Ibid.

11. Chan et al., "Reverse Transcription of Inserted DNA in a Monkey Gives Us Andi—Response," pp. 214–15.

12. Prather et al., "Transgenic Swine for Biomedicine and Agriculture," pp. 115–23.

13. Ibid.

14. Ibid.

15. "Outcry over Pinky and Yellowy," *Sky News*, http://www.sky.com/skynews/article/0,,15410-1032139,00.html [accessed December 9, 2003].

16. H. Pearson, "Engineered Pig Organs Survive in Monkeys," *Nature Science Update*, http://www.nature.com/nsu/031201/031201-9.html [accessed December 9, 2003].

17. R. S. Prather, "Cloning—Pigs Is Pigs," *Science* 289, no. 5486 (2000): 1886–87.

12. CAMELEONS, FLIP, FRET, FRAP, AND CAMGAROOS

1. J. Lippincott-Schwartz, N. Altan-Bonnet, and G. H. Patterson, "Photobleaching and Photoactivation: Following Protein Dynamics in Living Cells," *Nature Cell Biology* (2003): S7–14.

2. E. M. Judd, K. R. Ryan, W. E. Moerner, L. Shapiro, and H. H. McAdams, "Fluorescence Bleaching Reveals Asymmetric Compartment Formation Prior to Cell Division in Caulobacter," *Proceedings of the National Academy of Sciences of the United States of America* 100, no. 14 (2003): 8235–40.

3. Ibid.

4. Lippincott-Schwartz et al., "Photobleaching and Photoactivation: Following Protein Dynamics in Living Cells," pp. S7–14.

5. Ibid.

6. R. Ando, H. Hama, M. Yamamoto-Hino, H. Mizuno, and A. Miyawaki, "An Optical Marker Based on the UV-Induced Green-to-Red Photoconversion of a Fluorescent Protein," *Proceedings of the National Academy of Sciences of the United States of America* 99, no. 20 (2002): 12651–56.

7. Ibid.

8. D. Steele, "Cells Aglow," *Howard Hughes Medical Institute Bulletin* (Summer 2004): 22–26.

9. M. Damelin and P. A. Silver, "Mapping Interactions between Nuclear Transport Factors in Living Cells Reveals Pathways through the Nuclear Pore Complex," *Molecular Cell* 5, no. 1 (2000): 133–40.

10. K. Truong and M. Ikura, "The Use of FRET Imaging Microscopy to Detect Protein-Protein Interactions and Protein Conformational Changes in Vivo," *Current Opinion in Structural Biology* 11, no. 5 (2001): 573–78.

11. Y. Xu, D. W. Piston, and C. H. Johnson, "A Bioluminescence Resonance Energy Transfer (BRET) System: Application to Interacting Circadian Clock Proteins," *Proceedings of the National Academy of Sciences of the United States of America* 96, no. 1 (1999): 151–56.

12. P. Mas, P. F. Devlin, S. Panda, and S. A. Kay, "Functional Interaction of Phytochrome B and Cryptochrome 2," *Nature* 408, no. 6809 (2000): 207–11.

13. D. Prasher, R. O. McCann, and M. J. Cormier, "Cloning and Expression of the Cdna Coding for Aequorin, a Bioluminescent Calcium-Binding Protein," *Biochemical and Biophysical Research Communications* 126, no. 3 (1985): 1259–68.

14. N. Demaurex and M. Frieden, "Measurements of the Free Luminal Er Ca2+ Concentration with Targeted 'Cameleon' Fluorescent Proteins," *Cell Calcium* 34, no. 2 (2003): 109–19.

15. A. Miyawaki, J. Llopis, R. Heim, J. M. McCaffrey, J. A. Adams, M. Ikura, and R. Y. Tsien, "Fluorescent Indicators for Ca2+ Based on Green Fluorescent Proteins and Calmodulin," *Nature* 388 (1997): 882–87.

16. J. M. Kendall and S. Stubbs. "Fluorescent Proteins in Cellular Assays," *Journal of Clinical Ligand Assay* 25, no. 3 (2002): 280–92; R. Rudolf, M. Mongillo, R. Rizzuto, and T. Pozzan, "Looking Forward to Seeing Calcium," *Nature Reviews Molecular Cell Biology* 4, no. 7 (2003): 579–86.

17. G. S. Baird, D. A. Zacharias, and R. Y. Tsien, "Circular Permutation and Receptor Insertion within GFP," *Proceedings of the National Academy of Sciences of the United States of America* 96 (1999): 11241–46; T. Nagai, A. Sawano, E. S. Park, and A. Miyawaki, "Circularly Permuted Green Fluorescent Proteins Engineered to Sense Ca2+," *Proceedings of the National Academy of Sciences of the United States of America* 98, no. 6 (2001): 3197–202.

18. Steele, "Cells Aglow," pp. 22–26.

19. M. Fehr, W. B. Frommer, and S. Lalonde, "Visualization of Maltose Uptake in Living Yeast Cells by Fluorescent Nanosensors," *Proceedings of the National Academy of Sciences of the United States of America* 99, no. 15 (2002): 98 46–51; M. Stitt, "Imaging of Metabolites by Using a Fusion Protein between a Periplasmic Binding Protein and GFP Derivatives: From a Chimera to a View of Reality," *Proceedings of the National Academy of Sciences of the United States of America* 99, no. 15 (2002): 9614–16.

20. Fehr et al., "Visualization of Maltose Uptake in Living Yeast Cells by Fluorescent Nanosensors," pp. 9846–51.

21. I. Lager, M. Fehr, W. B. Frommer, and S. W. Lalonde, "Development of a Fluorescent Nanosensor for Ribose," *FEBS Letters* 553, nos. 1–2 (2003): 85–89.

13. CANCER

1. Karl S. Kruszelnicki, "Light of Life 2," *Great Moments in Science*, http://www.abc.net.au/science/k2/moments/s587114.htm [accessed October 31, 2003].

2. T. J. Sweeney, V. Mailander, A. A. Tucker, A. B. Olomu, W. S. Zhang, Y. A. Cao, R. S. Negrin, and C. H. Contag, "Visualizing the Kinetics of Tumor-Cell Clearance in Living Animals," *Proceedings of the National Academy of Sciences of the United States of America* 96, no. 21 (1999): 12044–49.

3. B. Laxman, D. E. Hall, M. S. Bhojani, D. A. Hamstra, T. L. Chenevert, B. D. Ross, and A. Rehemtulla, "Noninvasive Real-Time Imaging of Apoptosis,"

Proceedings of the National Academy of Sciences of the United States of America 99, no. 26 (2002): 16551–55.

4. Ibid.

5. Ibid.

6. K. Truong and M. Ikura, "The Use of FRET Imaging Microscopy to Detect Protein-Protein Interactions and Protein Conformational Changes in Vivo," *Current Opinion in Structural Biology* 11, no. 5 (October 2001): 573–78.

7. G. Caceres, Y. Z. U. Xiao, J. A. Jiao, R. Zankina, A. Aller, and P. Andreotti, "Imaging of Luciferase and GFP-Transfected Human Tumours in Nude Mice," *Luminescence* 18, no. 4 (2003): 218–23.

8. M. Chalfie, Y. Tu, G. Euskirchen, W. W. Ward, and D. C. Prasher, "Green Fluorescent Protein as a Marker for Gene Expression," *Science* 263 (1994): 802–805.

9. T. Chishima, Y. Miyagi, X. Wang, H. Yamaoka, H. Shimada, A. R. Moossa, and R. M. Hoffman, "Cancer Invasion and Micrometastasis Visualized in Live Tissue by Green Fluorescent Protein Expression," *Cancer Research* 57 (1997): 2042–47.

10. R. M. Hoffman, "In Vivo Imaging of Metastatic Cancer with Fluorescent Proteins," *Cell Death and Differentiation* 9, no. 8 (2002): 786–89.

11. M. Yang, E. Baranov, A. R. Moossa, S. Penman, and R. M. Hoffman, "Visualizing Gene Expression by Whole-Body Fluorescence Imaging," *Proceedings of the National Academy of Sciences of the United States of America* 97, no. 22 (2000): 12278–82.

12. M. Yang, E. Baranov, J. W. Wang, P. Jiang, X. Wang, F. X. Sun, M. Bouvet, A. R. Moossa, S. Penman, and R. M. Hoffman, "Direct External Imaging of Nascent Cancer, Tumor Progression, Angiogenesis, and Metastasis on Internal Organs in the Fluorescent Orthotopic Model," *Proceedings of the National Academy of Sciences of the United States of America* 99, no. 6 (2002): 3824–29.

13. Hoffman, "In Vivo Imaging of Metastatic Cancer with Fluorescent Proteins," pp. 786–89.

14. M. Yang, E. Baranov, X. M. Li, J. W. Wang, P. Jiang, L. Li, A. R. Moossa, S. Penman, and R. M. Hoffman, "Whole-Body and Intravital Optical Imaging of Angiogenesis in Orthotopically Implanted Tumors," *Proceedings of the National Academy of Sciences of the United States of America* 98, no. 5 (2001): 2616–21.

15. M. Yang, L. N. Li, P. Jiang, A. R. Moossa, S. Penman, and R. M. Hoffman, "Dual-Color Fluorescence Imaging Distinguishes Tumor Cells from Induced Host Angiogenic Vessels and Stromal Cells," *Proceedings of the National Academy of Sciences of the United States of America* 100, no. 24 (2003): 14259–62.

16. H. Zen, K. I. Nakashiro, S. Shintani, T. Sumida, T. Aramoto, and H. Hamakawa, "Detection of Circulating Cancer Cells in Human Oral Squamous Cell Carcinoma," *International Journal of Oncology* 23, no. 3 (2003): 605–10.

17. Y. A. Yu, T. Timiryasova, Q. Zhang, R. Beltz, and A. A. Szalay, "Optical Imaging: Bacteria, Viruses, and Mammalian Cells Encoding Light-Emitting Proteins Reveal the Locations of Primary Tumors and Metastases in Animals," *Analytical and Bioanalytical Chemistry* 377, no. 6 (2003): 964–72.

18. T. Theodossiou, J. S. Hothersall, E. A. Woods, K. Okkenhaug, J. Jacobson, and A. J. MacRobert, "Firefly Luciferin-Activated Rose Bengal: In Vitro Photodynamic Therapy by Intracellular Chemiluminescence in Transgenic Nih 3t3 Cells," *Cancer Research* 63, no. 8 (2003): 1818–21.

19. Ibid.

20. Ibid.

14. GLOWING GENES IN MEDICINE

1. Most of the medical background information presented here is derived from the Mayo Clinic Web site (http://www.mayoclinic.com), which is where I first go when I need information about medical conditions.

2. C. R. Cogle, S. M. Guthrie, R. C. Sanders, W. L. Allen, E. W. Scott, and B. E. Petersen, "An Overview of Stem Cell Research and Regulatory Issues," *Mayo Clinic Proceedings* 78, no. 8 (2003): 993–1003.

3. H. Wichterle, I. Lieberam, J. A. Porter, and T. M. Jessell, "Directed Differentiation of Embryonic Stem Cells into Motor Neurons," *Cell* 110, no. 3 (August 9, 2002): 385–97.

4. Ibid.

5. Susan Conova, "Custom-Made Motor Neurons," *In Vivo* 1, no. 15 (September 25, 2002): http://www.healthsciences.columbia.edu/news/in-vivo/Vol1 _Iss15_sept25_02/ [accessed November 3, 2003].

6. Cogle et al., "An Overview of Stem Cell Research and Regulatory Issues," pp. 993–1003.

7. A. J. Wagers, R. I. Sherwood, J. L. Christensen, and I. L. Weissman, "Little Evidence for Developmental Plasticity of Adult Hematopoietic Stem Cells," *Science* 297, no. 5590 (2002): 2256–59.

8. Ibid.

9. H. Beck, R. Voswinckel, S. Wagner, T. Ziegelhoeffer, T. Heil, A. Helisch, W. Schaper, T. Acker, A. K. Hatzopoulos, and K. H. Plate, "Participation of Bone Marrow–Derived Cells in Long-Term Repair Processes after Exper-

imental Stroke," *Journal of Cerebral Blood Flow and Metabolism* 23, no. 6 (2003): 709–17.

10. T. E. Schneider, C. Barland, A. M. Alex, M. L. Mancianti, Y. Lu, J. E. Cleaver, H. J. Lawrence, and R. Ghadially, "Measuring Stem Cell Frequency in Epidermis: A Quantitative in Vivo Functional Assay for Long-Term Repopulating Cells," *Proceedings of the National Academy of Sciences of the United States of America* 100, no. 20 (2003): 11412–17.

11. Ibid.

12. M. Hara, X. Y. Wang, T. Kawamura, V. P. Bindokas, R. F. Dizon, S. Y. Alcoser, M. A. Magnuson, and G. I. Bell, "Transgenic Mice with Green Fluorescent Protein-Labeled Pancreatic Beta-Cells," *American Journal of Physiology— Endocrinology and Metabolism* 284, no. 1 (2003): E177–83.

13. A. M. van der Sar, R. J. P. Musters, F. J. M. van Eeden, B. J. Appelmelk, C. M. Vandenbroucke-Grauls, and W. Bitter, "Zebrafish Embryos as a Model Host for the Real Time Analysis of *Salmonella Typhimurium* Infections," *Cellular Microbiology* 5, no. 9 (2003): 601–11.

14. Ibid.

15. Ibid.

16. G. D. Luker, J. P. Bardill, J. L. Prior, C. M. Pica, D. Piwnica-Worms, and D. A. Leib, "Noninvasive Bioluminescence Imaging of Herpes Simplex Virus Type 1 Infection and Therapy in Living Mice," *Journal of Virology* 76, no. 23 (2002): 12149–61.

17. G. D. Luker, J. L. Prior, J. L. Song, C. M. Pica, and D. A. Leib, "Bioluminescence Imaging Reveals Systemic Dissemination of Herpes Simplex Virus Type 1 in the Absence of Interferon Receptors," *Journal of Virology* 77, no. 20 (2003): 11082–93.

18. J. T. Trachtenberg, B. E. Chen, G. W. Knott, G. P. Feng, J. R. Sanes, E. Welker, and K. Svoboda, "Long-Term in Vivo Imaging of Experience-Dependent Synaptic Plasticity in Adult Cortex," *Nature* 420, no. 6917 (2002): 788–94.

19. Kathy Svitil, "Memory's Machine," *Discover* 24, no. 4 (2003).

20. G. S. Waldo, B. M. Standish, J. Berendzen, and T. C. Terwilliger, "Rapid Protein-Folding Assay Using Green Fluorescent Protein," *Nature Biotechnology* 17, no. 7 (1999): 691–95.

21. G. A. Caldwell, S. Cao, E. G. Sexton, C. C. Gelwix, J. P. Bevel, and K. A. Caldwell, "Suppression of Polyglutamine-Induced Protein Aggregation in *Caenorhabditis elegans* by Torsin Proteins," *Human Molecular Genetics* 12, no. 3 (2003): 307–19.

22. Joe Schwarcz, *That's the Way the Cookie Crumbles* (Toronto: ECW Press, 2002).

23. M. Friedman-Einat, L. Camoin, Z. Faltin, C. I. Rosenblum, V. Kaliouta, Y. Eshdat, and A. D. Strosberg, "Serum Leptin Activity in Obese and Lean Patients," *Regulatory Peptides* 111, nos. 1–3 (2003): 77–82.

24. Ibid.

25. H. Y. Liu, T. Kishi, A. G. Roseberry, X. L. Cai, C. E. Lee, J. M. Montez, J. M. Friedman, and J. K. Elmquist, "Transgenic Mice Expressing Green Fluorescent Protein under the Control of the Melanocortin-4 Receptor Promoter," *Journal of Neuroscience* 23, no. 18 (2003): 7143–54.

15. DEFENSE, SECURITY, AND BIOTERRORISM

1. Francine Jacobs, *Nature's Light: The Story of Bioluminscence* (New York: Morrow, 1974).

2. R. S. Burlage, "Green Fluorescent Bacteria for the Detection of Landmines in a Minefield," *Abstracts of the Second International Symposium on GFP* (San Diego, CA, 1999).

3. C. E. French, S. J. Rosser, G. J. Davies, S. Nicklin, and N. C. Bruce, "Biodegradation of Explosives by Transgenic Plants Expressing Pentaerythritol Tetranitrate Reductase," *Nature Biotechnology* 17, no. 5 (1999): 491–94.

4. T. H. Rider, M. S. Petrovick, F. E. Nargi, J. D. Harper, E. D. Schwoebel, R. H. Mathews, D. J. Blanchard, L. T. Bortolin, A. M. Young, J. Z. Chen, and M. A. Hollis, "A B Cell–Based Sensor for Rapid Identification of Pathogens," *Science* 301, no. 5630 (2003): 213–15.

5. R. Schuch, D. Nelson, and V. A. Fischetti, "A Bacteriolytic Agent That Detects and Kills *Bacillus Anthracis*," *Nature* 418, no. 6900 (2002): 884–89.

6. W. J. Ribot, G. Ruthel, B. J. Curry, and T. A. Hoover, "23rd Army Science Conference," (Orlando, FL, 2002).

7. E. W. Campion, "Disease and Suspicion after the Persian Gulf War," *New England Journal of Medicine* 335, no. 20 (1996): 1525–27.

8. Joe Schwarcz, *That's the Way the Cookie Crumbles* (Toronto: ECW Press, 2002), pp. 154–59.

9. Ibid.

10. A. J. Murk, P. E. G. Leonards, A. S. Bulder, A. S. Jonas, M. J. C. Rozemeijer, M. S. Denison, J. H. Koeman, and A. Brouwer, "The CALUX (Chemical-Activated Luciferase Expression) Assay Adapted and Validated for Measuring Tcdd Equivalents in Blood Plasma," *Environmental Toxicology and Chemistry* 16, no. 8 (1997): 1583–89.

11. Ibid.

16. LAST LIGHT

1. Clarissa Cheney, private communication.

2. R. Y. Tsien, "Imagining Imaging's Future," *Nature Cell Biology* (2003): SS16–21.

3. Ibid.

4. R. J. Kelso, M. Buszczak, A. T. Quinones, C. Castiblanco, S. Mazzalupo, and L. Cooley, "Flytrap, a Database Documenting a GFP Protein-Trap Insertion Screen in *Drosophila Melanogaster*," *Nucleic Acids Research* 32 (2004): D418–20.

INDEX